Horse

Animal
Series editor: Jonathan Burt

Horse

Elaine Walker

REAKTION BOOKS

In memory of Rooster and Joey

Published by
REAKTION BOOKS LTD
33 Great Sutton Street
London EC1V ODX, UK
www.reaktionbooks.co.uk

First published 2008
Reprinted 2013
Copyright © Elaine Walker 2008

Printed and bound in China by C&C Offset Printing Co., Ltd

British Library Cataloguing in Publication Data
Walker, Elaine
    Horse. – (Animal)
    1. Horses
    I. Title
    599.6'655

ISBN 978 1 86189 395 6

Contents

Introduction

Once I am in the saddle I never willingly dismount, for
whether well or ill, I feel better in that position.
Michel de Montaigne, *Essays*

The horse is a familiar animal. We do not think of it as rare or
exotic. Even in an urban setting, horses graze in paddocks on
the edges of many cities, control crowds at public events and
pull carriages for tourists. Horses appear in films, adverts and
logos and are among the few large animals that will interact
gently with a small child, so that most of us have memories of
offering handfuls of grass to an animal that loomed above us, its
alarming size compensated for by tickling whiskers and a willingness to stand while we patted its neck. The presence of
horses in our everyday lives is something we take for granted,
regardless of our own familiarity or knowledge of them beyond
the casual encounter.

Those who live alongside horses that are family pets or working companions may forget that not everyone has an
understanding of them. New owners looking for someone to
attach metal to the 'wood' on their horse's feet, or leaving a
week's worth of hay out and expecting the horse to ration itself,
are alarming to anyone who has grown up with horses around.
For those who know them only as an attractive sight in a field or
on a screen, the complexities of their nature can come as a surprise. When a non-horsey friend asked what I thought of the
fantasy elements of the film *The Horse Whisperer*, based on
Nicholas Evans's book, I had no idea what he meant. The tech-

Urban horse with
Leeds city centre
in the distance.

niques of working with the horse's natural instincts have become
so common in the past fifteen years that I was bewildered to find
he could think them entirely fictional. When I replied that the
film had no fantasy elements, he was equally bemused.

Like many children, I began the gradual pressure on my
parents to let me have my own horse at an early age. My father
recalls that, at three, I demanded to know why I couldn't have
one immediately. When he paused, in the process of shaving,

to explain that he couldn't afford it, I scanned the bathroom for evidence of needless extravagance and replied that it was no wonder, when he wasted so much money on soap and toothpaste. The priorities of a horse-lover rarely change and thirty years later I was fortunate in having a bank manager who bred show ponies. When I cautiously approached for a loan to buy the wonderful spotted Appaloosa horse I was yearning for, she smiled in complete understanding and said, 'Let's say it's for a conservatory.'

The kinship of such anecdotes links the like-minded down the centuries. Back in the 1650s William Cavendish, Marquis of Newcastle, exiled in Antwerp after the Royalist defeat at the battle of Marston Moor, lived so close to poverty that his wife had to pawn her jewellery so they could eat. Yet he recalls happily that, 'As poor as I was in those days, I made shift to buy, at several times, four Barbes, five Spanish horses, and many Dutch horses; all the most excellent Horses that could be.'[1] He was also freely disposed to giving horses as gifts, so his wife must have been in despair, especially when, in order to return to England, he had to leave her behind to secure his debts.

Tom Booker (Robert Redford) comforts Pilgrim in *The Horse Whisperer* (1998)

The world of the
pony-mad is
captured perfectly
in the popular
cartoons of
Norman Thelwell
(1923–2004).

HOW TO GET A PONY

Acquiring a pony is not quite as easy as it sounds . . .

Nearly four hundred years later, when William Holt met a
thin Arab horse called Trigger pulling a rag-and-bone cart, he
'stood for a moment, touched by a strange emotion, a sympa-
thetic yearning which stirred up fanciful ideas'. He spent the
evening watching the horse graze on a railway embankment
with express trains thundering past him, then, 'Next day I
bought him, the two-wheeled cart, the old iron and everything.'
He had no reason to buy a horse or, at least at first, any inten-

The family pony
from 1965 to 2007.

The decorative potential of horses: a composite horse from early 17th-century Golconda.

tion of riding him. At first he thought he had adopted the horse 'as one does a stray dog for the pleasure of making him happy', but soon realized that 'It was he who rescued me.'[2] Before long, at the age of 67, Holt set out with his new friend and together they travelled 9,000 miles through western Europe.

Yet stories of the abuse of horses fill hours of animal rescue programmes on television and neglect, from ignorance to wilful cruelty, seems as commonplace as the much-loved horse that is a family member. It seems that our interaction with horses can bring out the best or the worst in us. But, at whatever level we relate to the horse, it is so closely linked with human development that without it our own history would be completely different.

Before the days of the car, the horse was a vital means of travel. Before the tractor, it was essential to agriculture. Horses have pulled barges, turned waterwheels, carried humans thousands upon thousands of miles in peace and war, across continents and landscapes they were not born to. They have been left behind when wars were over, or resources ran out, to survive as best they could and have often done that very well, constantly adapting to their surroundings. They have also died from the wounds of swords, arrows and explosives, from starvation, exposure or exhaustion. The potential offered by a horse goes far beyond strength or speed, however, and its true value comes in its willingness to interact companionably with humans, whom its very nature teaches it first to fear and then to follow.

It is the similarities and differences in our natures that draw us together, making this companionship possible. A horse is a

An engraving from a collection of flora and fauna by Wenceslaus Hollar, 1663.

prey animal, depending for safety on its herd-leader and the ability to fly from danger. In the wild, a herd is guarded by a stallion and led by a matriarchal mare. The domestic horse still understands and needs confident leadership and to know its place in the herd. Veteran horseback traveller Jeremy James bought horses for his journey from southern Bulgaria to Romania en route. Then he rode these strangers to him and each other over mountains, through rivers, under fire over an army gunnery range and, on one occasion, through 'this impossible pedestrian precinct . . . with trains screaming past'. Though he says, 'I can tell you it's no joke going through somewhere like Paddington underground at rush-hour with a pair of horses. Never do it. I don't know why no one got killed, I don't know why we made it,' the clues to his success are evident. Part way through their journey, stranded in a rainstorm on a precipitous mountainside, lost and hungry in the pathless loneliness of a Transylvanian forest, he made his horses, Karo and Puşa, a promise that they would 'never be afraid again': 'I promised them that I would take them to my home. I promised them that they wouldn't have to walk any more, that I would take their shoes off and they could play and graze in green fields that look over the changing colours of Brecon, where the streams lie, south facing, in the eye of the sun.'[3]

Horses may not understand what we say to them, but they understand how we say it and they understand when they can trust us. On that basis, they will follow a human through fire, or an underground station. A bullying, cowardly or uncertain leader, however, will inspire only an uneasy and perhaps resentful horse, with an unhappy spiralling decline of mistrust and danger to follow for both parties.

As humans, we understand these instincts for reliable leaders and trusting relationships. While much of our society is

based on concepts of hierarchy, the human family is not naturally based on dominance, as we are often led to believe, but essentially on trust in the wisdom and confidence of the parent. The guarding father and guiding mother are not uncommon figures to human children, even as our own social norms change and mutate with time. In this we are like the horse, looking to be led and kept safe until such a time as we can lead and make safe our own young. Among our peers, we establish relationships based on levels of trust and familiarity. When we are forced to submit through fear, we are likely to become cowed or rebellious. Horses are the same. While the image of the dominant stallion is a traditional one, the understanding of the role of the lead mare and the concept of the passive leader brings a shift that aligns us more with a true understanding of our relationship with the horse.

Follow-my-leader is a way of life for wild horses.

Study of a horse's head, by Antonio Pisanello, *c.* 1430–40.

The passive leader is the ultimate role-model, who leads by confidence and example, not dominance, and while the weak or afraid may submit before the dominant leader, it is the passive leader, the one who simply 'is' leader, who inspires faith. This applies equally well to a human or a horse. Small wonder that we find the possibility of a relationship with a horse, unlike the

Controlling the horse as a symbol of human power in a statue in Fougères, Brittany.

grateful adoration of a dog, or the cool independence of a cat, essentially one of finding the right place in which to interact co-operatively for mutual safety in whatever situation the horse–human pair finds themselves.

This relationship was once essential to survival in many civilizations across the world. Then came the time when these subtleties were worn away as war and commerce lacked sufficient time for the nuances of interaction to be fully worked out. Fierce metal bits and leather straps, coercion instead of co-operation, offer quick control and, because the horse is a prey animal and the human a predator, this is accepted at a different level. The aspect of human nature that seeks to dominate can find satisfaction in the strapped-down, fiercely bitted 'broken' horse with heaving flanks and bleeding sides. By a strange irony, as the horse becomes less vital to modern human expansion, its treatment has become subtle once more, and huge amounts of research are undertaken today into a more profound understanding of an animal whose essential nature is gentle and wary, relaxed but ready to run.

The importance of the horse in our history is written into our language in sayings with a long past, still spoken today without much thought. While the meaning of 'shutting the door after the horse has bolted' is obvious, that a 'cavalier attitude' derives from the word for a horseman, or that 'looking a gift horse in the mouth', refers to telling a horse's age by looking at its teeth would have had a different relevance in the days when those activities were part of daily life. That this last-mentioned saying was proverbial in Latin as early as the fourth or fifth century illustrates the long interweaving of horse references through history and across cultures. Today many are simply metaphors we understand due to their familiarity, rather than a reflection of our experience. Once, being told to 'get back in the saddle' did not just mean get over the setback and have another go: it literally meant to get back on the horse you had fallen from before reluctance set in, a common practice still when teaching a novice to ride. Every time we casually use sayings like these we take an unconscious step back to the time when the horse was an essential part of everyday life.

Our relationship with the horse has changed completely since the Industrial Revolution. From the beginnings of civilization, the horse has played a part, first as a food source, then as a means of better hunting. Once the possibility of riding the horse became clear, travel and commerce opened up for humans in ways we have today pushed beyond the horse's abilities into the age of technology. Yet we still measure mechanical capability in 'horsepower', a term coined in the late eighteenth century by James Watt, whose discoveries, somewhat ironically, took horses out of the equation. The horse has served as a means of transit in many areas and even today, in countries such as Iceland, the surefooted abilities of native horses can make riding the safest and most efficient means of travelling across difficult landscapes.

Yet there is something more to our relationship with the horse than convenience. A horse is an animal with which humans can make a relationship that impacts upon our ability to handle and understand it. While it does not have the same responses as a dog, it can form just as close a bond with a regular handler. Feeling insecure alone, in return for human company it shows initiative and an ability to learn. Ways of training horses have changed over the years, often reflecting changes in humans' understanding of themselves, rather than of horses.

Humans are drawn to animals with large eyes, as we are to babies. Study of a horse by Daniéle Da Meda.

There is a Buddhist saying that 'it is your mind that creates this world', and the human tendency to impose our current viewpoint upon our understanding of the world makes that creation one of continual change. What to one handler may be a recalcitrant horse that deserves to be beaten into submission will to another be a nervous prey animal that needs reassurance.

Lucy Rees tells of a horse that had been ridden by a series of 'bronc-busters' and written off as uncontrollable, saying, 'I learned this story with some surprise, for I had just ridden her to a new field in a halter'. After several months' work to reassure the horse that being ridden did not necessarily involve being frightened, Lucy returned the mare to her owners, advising that the horse should be ridden quietly and gently, preferably by a woman. They 'put a cowboy on her just to make sure' and after that she 'was never ridden again'.[4]

The handler may sometimes even be the same person whose views change but the horse will remain the same, motivated by instinct and habit. The paradox of the horse is that by any measure we choose it is far too strong for us to handle. A frightened horse can easily kill an adult without any malicious intent, simply by an untimely swing of its weighty head, yet stories of a horse standing immovable until someone notices the small child hugging its back leg are many.

Through this book, I will trace just a few of the countless paths humans and horses have travelled together, through changes in culture and technology and for the purposes of discovery, war and pleasure. My own long experience with horses will be my starting point, along with the experience of others across the centuries, expressed through art, writing and modern media, for exploring the ways in which horses and humans form and inform one another.

1 Eohippus to Equus

Wild Horse breathed on the Woman's feet and said, 'O my
Mistress and Wife of my Master, I will be your servant for the
sake of the wonderful grass.'
Rudyard Kipling, *The Just So Stories*

Deep in the rich tropical forests of prehistory, padding its cautious way on long-nailed paws, a small deer-like animal browsed on low shrubs and the rushy margins of marshland pools. This little creature was not only the ancestor of the horse, it was also something of a red herring. When its remains were discovered in the 1870s by American palaeontologist, Othniel C. Marsh, he came to the conclusion that this was the missing link to an understanding of the development of the modern horse.

Some years earlier, in the 1840s, fossil remains from the same genus were found in London by Sir Richard Owen, who named it *Hyracotherium*, a name considered today to be more accurate and less misleading. However, with the discovery of 'horse' fossils in America, long considered to have no indigenous horse breeds, this small creature was named twice, and the second name, *Eohippus* or 'dawn horse', while more romantic, led to a misconception.

The first part of this was that the *Eohippus* was an early horse, when it was actually also the ancestor of many other animals, including the tapir and the rhinoceros. The second was that it was not only an early but a primitive, as yet unfinished, model of the horse we know today, something along the lines of a work-in-progress. The horse we know today would not have survived long among the swampy tropics of the Eocene era, while the

The skeleton of a *Hyracotherium*.

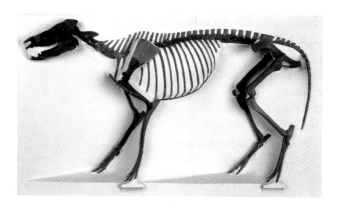

Hyracotherium was perfect for its own time and changed little over 20 million years. It had legs which would flex and rotate to allow a wide range of movement and had developed for a soft-soiled environment, with four toes on each front foot and three on each hind. It walked not on hooves but on strong nails and a 'heel' in the form of a hard pad. The vestigial remains of this pad are thought to be found today on the horse in the ergot, a small nub of hoof-type material on the point of the fetlock. It had a small brain and eyes on the front of its head, not the sides like a horse, illustrating that it was a creature needing forward, rather than peripheral, vision. This suited its dense and sheltered environment, while a coat pattern of soft spots would have been its most likely colouring, providing camouflage among the leafy shade of its habitat.

Skeletons of the *Hyracotherium* have been found in America and Europe and suggest that its size ranged from around 25 centimetres (10 inches) to maybe twice that at the shoulder, and that it had a similar wide variety of shape and size across its range of locations. It seems likely to have spread across the land bridges that once joined the continents of America, Asia and Europe and remained at this evolutionary stage for several million years. It is easy to forget, when comparing the animals of prehistory with those familiar to us today, that the theory of evolution does not suppose the linear movement of any organic form towards a perfect end. While the changes over millennia may be seen as developments, this does not suggest that the original form was imperfect, but rather provides evidence of an ability to adapt that ensured that certain branches of the family survived and others died out.

Mid-way on the long journey between *Hyracotherium* and the hoofed *Equus* of today, during the Oligocene period, came the *Mesohippus*, by way of the *Orohippus*, the *Epihippus* and some 20 million years. Still small at 45 centimetres (18 inches) but starting to develop teeth that could manage a greater range of foliage, and feet that could cope with harder and more varied terrain, the *Mesohippus* was the natural response to a cooling climate. As abundant grasslands spilled across the surfaces of the changing landscape to take the place of the receding forests, the teeth of this animal became larger and more durable. Also, the longer legs suggest that greater speed had become important, while with longer teeth came a larger skull with more lateral vision, all as adaptation to the more open habitat. Slowly, the *Mesohippus* was changing from a shy forest creature, relying upon concealment for safety, to a flight animal whose speed and agility were its best means of defence.

Artist's reconstruction of the world of the *Mesohippus*.

As the earth shifted gradually towards a more temperate climate over the next several million years, the little creature grew taller, with its toes becoming vestigial. It started to take its weight on a central toe, which gave it a longer stride, while its teeth became more recognizably those of a grazing animal. The *Miohippus* overlapped with the *Mesohippus* for around 4 million years and fossil evidence suggests that the different species, and other variations that had split off from the *Mesohippus*, coexisted. Such families of related animals, 'clades' to the palaeontologist, illustrate the diversification of species in order to take advantages of various changes in their surroundings. This means that while they may follow a general tendency, such as, in the example of the *Mesohippus* and its relations, having a larger brain than most herbivores, they each also develop traits of specific use to their own location and climate. The negative aspect of this is that, should those specifics change, a complete branch of the family may be wiped out quite suddenly. Those who survive may simply be in the right place at the right time, rather than having any superior qualities.

While a simple overview might suggest that the horse evolved from a small animal into a large one, palaeontologists find that the different species grew both larger and smaller than their progenitors, again to adapt and survive. In *The Great Evolution Mystery*, Gordon Rattray Taylor concludes that

> the fact is that the line from Eohippus to Equus is very erratic. It is alleged to show a continual increase in size, but the truth is that some variants were smaller than Eohippus, not larger. Specimens from different sources can be brought together in a convincing-looking sequence, but there is no evidence that they were actually ranged in this order in time.[1]

Therefore, although Othniel Marsh's belief that 'The line of descent appears to have been direct'[2] was presented as fact for almost a century, today it is widely accepted that the evolutionary picture is far more complex.

Around 6 million years ago, emerging from the changes in teeth, body size, leg length and facial structure that had occurred during the Miocene era, came animals looking not unlike a mule that became the source not only of the modern horse, but also of zebras and donkeys in all their variations. After a successful few million years for several three-toed plains grazing species, the line that led to the horses of today developed with the formation of side ligaments that supported the central toe and eventually superseded the two additional toes. This change to a single hard toe, with tendons that allow the gather and release of energy in a springing movement, meant that *Pliohippus*, along with the wonderfully named *Astrohippus* and *Dinohippus*, were the earliest true relations of *Equus*, the genus of all modern species.

The first *Equus caballus*, the ancestor of today's horse, is thought to have descended from the *Pliohippus* relatively

Cave painting of hunters from Indian Creek, Utah.

A palaeolithic
painting at
Lascaux, France.

recently, around one million years ago. Considering the rate of
change over the previous 50 million years, this represents something of a rapid development. The last period of glaciation ended
approximately 10,000 years ago and as this final Ice Age receded
the more clearly defined forms of *Equus* established themselves:
horses in Europe and Western Asia, asses and zebras in the north
and south of Africa, and onagers in the Middle East. While it
is commonly accepted that horses were introduced into the
Americas in the sixteenth century by Spanish conquistadors,
their ancestors had roamed the continent up until 8,000 years
earlier before becoming completely extinct there for reasons not
fully understood. Climate change, food shortage and predators
have been put forward as possibilities, though there is another

consideration, which, if true, was perhaps something of an omen for the horse. Excavations of prehistoric cooking sites indicate that horses had become an important part of the human diet, leading to growing consideration of the impact of human predation on the early horse population. Archaeological evidence suggests that within 2,000 years of the arrival of man in North America, horses, along with other herd animals including camels, were gone. Archaeologist and ecologist Paul Martin concludes that 'This extinction was postglacial in time and affected in the main the larger animals. The principal factor isolated as cause is the appearance of man.'[3]

The wide distribution of early horses and the way in which they have always survived by adapting to their environment are partly responsible for their impact upon human development. At its simplest level this can be seen by observing any modern horse. There is a tendency today to provide rugs and winter shelter for horses, which makes them more comfortable but inhibits their natural capacity for adaptation. Left to themselves, horses

The ample figure of this horse indicates the successful career of his rider, a red-coated official, in a coloured ink drawing by Zhao Mengfu, c. 1296.

will grow a winter coat and shed it depending on location. Of course, many would die in very severe weather or a time of food shortage, so as the horse developed, its survival became dependent upon its strength and resilience. The animals that could adapt and cope with hardship most readily forged their way through the years, developing the core features that determined their future. Different types of horse evolved to suit the climates they inhabited, so horses in hot, dry climates developed slender limbs and fine skin with veins prominent on the surface to aid cooling. Cold-climate horses developed as small and solid, with coats which readily grow dense in the winter and legs which develop long hair, known as 'feather', especially round the fetlocks, the horse's heel, to shed water.

Among many types a spotted horse will occasionally appear. Examples may be found in images as old as the cave paintings of Peche-Merle in southern France which date to around 20,000 BC, as well as Etruscan tombs in Italy and burial sites at Hallstatt in Austria dating from around 800 BC. They are found decorating weapons of the Celts of northern Europe and the nomadic steppe horsemen of south-western Russia. They also appear in legends, such as those of the Persian hero, Rustam, whose horse, Rakhsh, is often depicted with a red spotted coat, and are particularly popular in early Persian and Moghul art. From the fifteenth century onwards, the selective breeding of spotted horses became prevalent in very different cultures, with the most well known being the Knabstrup of Denmark and the Appaloosa of the Nez Perce tribe of north-eastern America, both descended from early Spanish strains. While these have developed into clearly defined breeds, the spotted gene seems to spring up here and there across time and location to recall that, like people, horses have a common ancestry and were originally separated by time and distance rather than blood.

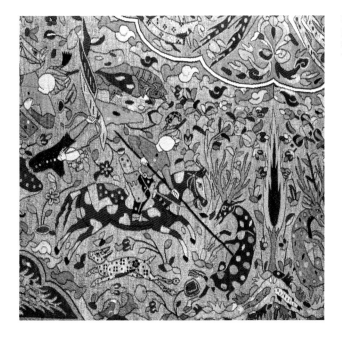

Spotted horses and prey in a Persian-style carpet from 17th-century India.

The human desire for order and reliability led to selective breeding, but the raw materials with which humans had to work were already established into particular types suited to the environment in which they lived. Most of these ancestors of modern horse breeds may have gone on to be shaped by human intervention but they were established by one of the key features that make a horse desirable for domestic use in the first place: its ability to adapt to changing conditions in order to survive. Today, while many examples of this tendency to adapt are evident, most of those horses considered to be native to a particular place are nevertheless the result of organized breeding programmes that sometimes go back centuries. However, these are based on promoting and developing the natural

A 16th-century spotted Japanese horse tries to kick his way out of his stable. Note the girth suspended from the ceiling of the stable to allow him some freedom of movement.

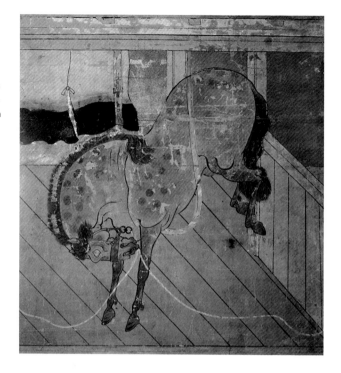

characteristic features originally developed to cope with a particular climate or terrain. A useful example is found in the Welsh Mountain Pony, which is believed to have been roaming the steep landscapes of Wales before the arrival of the Romans. Small, hardy and intelligent, it developed to survive on sparse feed, rough and dangerous mountain terrain and through severe weather. A strong suggestion of Arab blood in the elegance of its head and shape of its ears may derive from the days of Roman occupation when Arab-type horses from the Roman campaigns in Africa were introduced into Wales. A long history of uses from baggage carrier to pit pony to the popular child's

riding pony of today have all derived from its essential characteristics, honed long ago by its environment.

The Welsh Mountain Pony has a long history and modern popularity.

Many of the horses described as 'wild' today are actually feral
horses – those descended from escaped or abandoned domestic
stock, such as the mustangs of America and the Australian
Brumby. These horses nevertheless illustrate the same adaptability that has enabled their survival without the protection of
domestication, developing the ability to thrive in poor conditions
while growing hardy and resilient. The presence of so many ill-
nesses associated with domestication, such as laminitis, a
metabolic disease similar to gout in humans, suggest too that the
cosseted lifestyles of many modern horses with rich grazing and
high energy feeds can be at odds with their nature as survivors.

Only one genuine species of wild horse survives and that is
the *Equus ferus przewalskii*, the Asiatic or Mongolian wild horse,

rediscovered in the late nineteenth century. Commonly known as the Przewalski's Horse, it is named for the Russian military surveyor and naturalist Colonel Nicolai Przewalski, who located the last of the wild herds of horses in 1879. Known locally as the Taki, they roamed the Tachin Schah, Mountains of the Yellow Horse, on the edge of the Gobi desert. Although wild horses had been known previously in this area, they were widely believed to have died out or been hunted to extinction for food and skins by the Kirghiz people. Further animals were discovered by a filial hunting partnership, the Grum-Grzhimailo brothers, and, with the realization that small herds still survived, efforts were made to save this last surviving link with the domestic horse's wild past.

While keeping rare animals was fashionable among the elite of Europe, the Zoological Museum of the Academy of Science in St Petersburg showed special interest in a species of horse that could genuinely be defined as 'wild'. They were also of interest to early conservationist-collectors, most notably Baron von Falz-Fein of Askania-Nova in southern Russia, and Herbrand, Duke

The Grum-Grzhimailo brothers with one of the last wild Przewalski's Horses.

of Bedford, of Woburn Abbey in England, who both had nature reserves on their private estates. Between 1897 and 1902 several expeditions set out to capture breeding stock from the surviving herds. During the most successful, initiated by Falz-Fein in 1901, 52 foals were caught using 'uraks', long poles with loops on the end, having been run down with faster horses and separated from their mothers and lead stallion, which, ironically, were shot to facilitate the capture. The expedition had not allowed for such a large catch and could not feed them all, so only 26 reached Europe alive, having made a 50-day journey on foot, then travelled by train and ship with domestic mares as foster-mothers.

A dispute arose when Carl Hagenbeck, a well-known animal dealer from Hamburg, was commissioned to buy the foals for the Duke of Bedford from under Baron von Falz-Fein's nose, although there are conflicting accounts of how this came about. After this rocky start, two further expeditions in 1902 and 1903 caught several more horses and over the next ten years the surviving animals were sold to zoos and private parks in Europe and America. The journey from their native country to their new homes was gruelling and one pair was rejected by the New York Zoo because of their poor condition. Hagenbeck had such difficulty finding a buyer for another horse that it was eighteen years old before he managed to sell it. There were few further captures and at length Przewalski's Horses became more established in captivity than in the wild, so that by 1969 they were feared to have died out altogether in their native habitat. Inbreeding and other problems of captivity affected the captive population so in 1977 the Foundation for the Preservation and Protection of the Przewalski's Horse was formed to start a stud-book to keep track of breeding and to work towards reintroducing the horses into the wild. By 1998, 60 Przewalski's Horses were living back on their native steppes, with a nature reserve set up to protect them.

The Przewalski's Horse has 66 chromosomes, rather than the 64 of the domesticated horse, and remains resistant to contact with humans even when kept in captivity. Like the zebra, it has a tendency towards aggression, unusual in *Equus caballus* without significant provocation. It is therefore probably not directly ancestral to modern breeds but instead is a closely related separate species. In appearance it is very similar to the images of hunted horses shown in the earliest of cave paintings.

Wild, feral and domesticated horses are not only different in their status of belonging or not belonging to humans. Once a wild animal becomes domesticated, over the generations it changes, as do its human keepers, through the shared relationship. A key aspect of the suitability of an animal for domestication is its willingness to breed while in captivity and also to interact easily with humans, usually through routine and an acceptance of the human as leader. It is the willingness to accept all the benefits of domestication while refusing to follow any leadership but its own that makes a cat far more complex to train than a dog or a horse, both of which are co-dependent animals. The horse's instincts are those of a prey animal and its flight response to danger is the strongest and, not surprisingly, the one humans find the most disconcerting in a domesticated situation. A horse does not naturally respond well to being tightly enclosed, approached by large and noisy unfamiliar objects, having anything on its back or its feet restricted. All of these situations suggest to it that it is in the presence of a predator, and yet it learns to be stabled, to go out in traffic, to be ridden and to have its feet handled regularly. Not only does it accept this 'predatory' behaviour but, depending upon its past experience and individual character, can do so without any concern at all.

It is partly this adaptability and partly its natural social structures that make the horse suitable for domestication.

An unhappy horse resists being tied while his companions graze freely in this image from China, which could be read as symbolic of the desire for liberty in every living creature

The wealth of nations: horses, giraffes and dogs, amongst other animals presented as tributes to the Egyptian Emperor Rekhmire by ambassadors from Nubia and Syria, c. 1425 BC.

Relatively few animals are domesticated and while experiments in riding zebras and hybrid equine crosses often have some level of success, there is always an element of unpredictability involved. The domesticated horse has been domesticated for a very long time and at the most simple level this makes a great difference. How that came about in the first place can be conjectured through the common understanding of the horse today.

As a herd animal, a horse is drawn to seek safety in numbers and to follow confident leadership. While its instincts would make it see humans as predators, if it were not pursued it would possibly also, over time, begin to see safety in human

Training a stallion
in an Etruscan
fresco painting.

settlements, from which other predators were kept away.
Natural curiosity and sociability on both sides might also have
played a part in initiating a change in the relationship between
horse and human, and once horses began to be tamed, ridden
and driven the future changed irrevocably for both parties.
With riding and pack-horses, travel became more manageable,
while a harnessed horse made great strength available for tow-
ing or carrying. The search for new land and resources, the
ability to fly from danger, as well as a ready source of hides,
meat, milk, and dung for building material and burning as
fuel, all became readily available once the horse became a
domestic animal.

Large depositories of bones in sites such as Solutré and
Lascaux in France suggest that initially the horse was simply a
source of meat and useful additional materials, such as bone
and skin. However, ultimately the attributes of speed and power
offered more use to humans and this impacted upon the keeping, breeding and handling of horses. Early depictions, such as
those in Santander, Spain, which date from around 13,000 BC,
show horses being hunted, and a successful hunt requires an

understanding of the animal. The first attempts to domesticate the horse most likely came from insights into its nature gained through hunting.

The horse was not, however, among the first domesticated animals. Sheep and cattle farming and a relationship with the dog were well established by 4,000 BC, by which time it is known that horses were domesticated. Reindeer were used to pull sledges and possibly ridden in northern Europe by 5,000 BC and cattle were used as draught animals before horses. However, once the potential of the horse was realized, its domestication spread westward from the grass steppes of Central Asia into western and central Europe, the Caucasus, and then into Arabia and China. Unlike the reindeer the horse was non-migratory, so it could be moved wherever its

Stylized horse from a Chinese painting, c. 535–56.

nomadic keepers chose. Unlike cattle it had great speed, while unlike the camel it was sociable.

While domestication of many animals was successful, it was the horse that for the next 4,000 years would offer the speed, strength and intelligence upon which the spread of human civilization could be founded. However, these qualities require a different approach to that suitable for cattle or sheep and horses do not appreciate the pack relationship that dogs respond to. For the sake of the wonderful grass promised him three times a day, Kipling's Wild Horse 'bent his wild head and the Woman slipped the plaited-hide halter over it'.[4] While it seems unlikely that the first domestic horses kept their part of the bargain quite this readily, that they would see its benefits is more so. In return for domestication the horse loses its freedom. But it loses too the constant anxiety of the prey animal, the relentless walking from one grazing ground to another and seeing its young starve to death during a hard winter, or being left behind to be eaten alive because of an injury. Horses appreciate safety and if humans could offer that, along with plentiful food, then the possibility for a willing partnership could emerge.

Archaeological evidence suggests that domestication of the horse occurred in several places at different times, rather than in a single place that then influenced others. In 1993 a collaboration between the Carnegie Museum of Natural History, the University of North Kazakhstan and the North Kazakhstan History Museum began to excavate a settlement located on in the Iman-Burluk River and dating from 3,600–2,300 BC. Known as Botai, the settlement has revealed a culture in which horses were central for food but may also have been domesticated. Artefacts suggesting bridle parts and skeletal remains of horses showing wear on the teeth support this theory, alongside evidence such as the presence of horse remains outside a natural range and features of size and proportion which suggest controlled breeding. However, interpretation of these findings varies and definitive dates for the earliest domestication of horses are unlikely to emerge due to the perishable nature of most of the equipment, such as rope and leather, used in handling. Any culture that enjoyed eating horses would soon realize that keeping them like cattle made their meat, milk and hides more readily available than hunting them in the wild. The sociable nature of the horse would make taming captive foals, or even older horses, possible with patience. Once its potential for carrying and draught became apparent, riding would almost inevitably follow. Pictorial as well as archaeological evidence from Pazyryk, in the Altai mountains of western Siberia, suggests that the Scythians had a history of horse-handling that stretched back to at least 3,000 BC, while archaeological evidence suggests that horses were imported from Scandinavia into Britain during the Bronze Age and were widely ridden and driven by the Viking and Baltic peoples from around the fifth to the eleventh centuries AD, while the etymology of Lithuanian, a language considered remarkable for its unchanging form, includes words derived from specific references to domesticated mares. The characteristic carved monuments of the Picts, the Celtic people of northern Britain, dating from the fifth to the ninth centuries AD illustrate that several types of horses were kept for purposes from riding to agriculture. Alongside evidence of grain production and training, it seems clear that, despite the difficulty in precisely dating some archaeological evidence, the handling and management of horses were well established and organized across a wide area during that time.

From the earliest days of interaction humans and horses have had a profound impact upon one another's lives and essential development. In making the journey from *Eohippus* to *Equus*, the horse became a creature that impressed humans not only by the practical benefits it could bring, but also by the level of relationship possible and the aesthetic impact upon human sensitivities of its beauty and speed. The day its role as a source of food and materials mutated to that of a servant or companion, irrevocable changes took place which would shape the future of both horse and human for thousands of years to come.

2 Pegasus, Epona and Demeter's Foals

Far back, far back in our dark soul the horse prances.
D. H. Lawrence, *Apocalypse*

The horse is an enduring image for many cultures, with its symbolic resonance often moving across continents, languages and eras. Myths and legends weave the horse into the fabric of storytelling in all its forms, while themes such as the passage between realms and horses as human advisors emerge time and again from divergent sources. The strong links between horse and human predate its domestication and show for how long a journey of mutual influence has been travelled, a journey that remains ongoing across the world.

One of the most familiar and recurring images is that of the winged horse Pegasus. In Greek mythology Pegasus was fathered by the sea-god Poseidon in the form of a horse upon the serpent-haired Gorgon Medusa. In some versions of the story the snakes that served Medusa for hair were inflicted on her by Athene who was outraged that the seduction should have taken place in her temple. Medusa is best known for her attempts to overcome Perseus by turning him to stone with the glare of her eyes. When Perseus defeated her, he cut off her head and Pegasus sprang from her body.

Several legends about the role of Pegasus exist and either the thunder of his galloping or a single blow of his hoof are said to have caused the Hippocrene Well to spring forth upon the slopes of Mount Helicon, long associated with poetic inspiration.

St George and the Dragon appear in many forms, but particularly in the iconography of the Russian Orthodox Church. This image dates from the 16th century.

Pegasus was given to Bellerophon for killing the Chimera, and the hero tamed him with the aid of a magic golden bridle, but then tried to fly his horse to the heavens. He was punished for this arrogance by Zeus, who sent a fly to sting Pegasus, making him throw his rider to the ground. Pegasus continued his transilience to Mount Olympus alone, where he carried lighting bolts for

Pegasus and Bellerophon fight the Chimera, from an emblem book of 1584. Note the similarities with St George and the dragon.

Zeus, and also became a constellation (which may be seen at its best in October) thus writing the legend into the night sky.

The association of Pegasus with inspiration and aspiration has remained constant and been used to symbolize just about every sort of enterprise from theatres to software companies. Pegasus was the shoulder insignia of the British Airborne Forces in the Second World War and, at the outset of the Normandy Invasion in June 1944, the bridge over the Caen Canal near the town of Ouistreham was renamed the Pegasus Bridge after a successful mission by paratroops to seize it for the Allies.

On the roof of the Magnolia Hotel in Dallas, Texas, a huge red porcelain-covered double image of Pegasus, 10.5 by 15 metres (35 by 50 feet) in size, has become linked with the city as an image of motivation. Originally erected in 1932 to celebrate the first meeting in Dallas of the American Petroleum Institute at the building, then the headquarters of Magnolia Oil, the sign could be seen for over 80 km (50 miles) when lit up and rotating. It fell gradually into disrepair and was eventually switched off in 1997, but was completely rebuilt in time for the millennium, when it was relit at midnight to welcome in the year 2000.

In Caravaggio's *Conversion of St Paul on the Way to Damascus*, c. 1601, the white patch on the skewbald horse's shoulder focuses the blinding light of God onto the prone figure of Saul.

Pegasus is only one of many mythical horses that have crossed cultural boundaries by recurring time and again in contexts linked only by the presence of the horse itself. A magic flying horse, which obeys instructions given by turning its ears, appears in *The Arabian Nights*, *The Canterbury Tales* and a traditional Jewish folktale thought to have originated in Afghanistan. This horse, made variously of wood, gold or flesh, is, like Pegasus, not so much a character in itself as symbol of human ingenuity and is conveniently forgotten when it is no longer useful to the tale:

> The prince was taken with a great curiosity to know more about it, so he mounted his magic horse, turned its left ear and flew directly onto the tower, landing on its roof. There he left the horse, and climbing down from the roof he found himself outside of a window. And when he looked within he saw in the brilliantly lit room a beautiful girl who was sound asleep.[1]

Obviously, a horse which can be left untended on a roof during a romantic tryst is invaluable but magical relations of the common horse often offer all of its qualities with none of its needs. An awareness of the deeper value of the animal comes not only in the horse as a source of inspiration, common in many cultures, but also in the link between earth and the heavens embodied in the flying horse and echoed in connections between horses and the spirit world.

These may come either through flight, transcending the earth, or through liminal settings such as cave or water dwelling places, where the horse becomes the guardian of the passage to the spirit underworld. While, as in the case of Pegasus, the provenance of these ideas can sometimes be traced, perhaps more interesting are the many horse myths and legends which seem to

Pegasus flies atop the Magnolia Hotel in Dallas, Texas.

Sleipnir, the eight-legged horse of Odin, in a funerary stone from 8th-century Sweden.

occur without connection across a diverse range of contexts. The horse as a sacred key to those aspects of life beyond the everyday understanding, to which humans turn for revelation and consolation, arises world-wide. Pictograph images of horses may be found on oracle bones which date to as far back as the Shang Dynasty of 1766–1045 BC in China. Thought to represent both actual horses and symbolic associations with the horse, such as masculine energy, they became part of the written language that developed from these early records of the messages of divination.

Through the trade along the silk route, the cross-pollination of cultures brought new images of the horse and took those familiar in the East beyond their native boundaries. The 'War God's Horse Song', from the original Navaho of Tall Kia Ah'ni, interpreted for an English speaking audience by Loms Watchman, is an example of the way in which cultures relate to one another through a shared inspiration:

I am the Turquoise Woman's son.
On top of Belted Mountain
Beautiful horses – slim like a weasel!
My horse has a hoof like striped agate;
His fetlock is like a fine eagle-plume;
His legs are like quick lightning.
My horse's body is like an eagle-plumed arrow;
My horse has a tail like a trailing black cloud.[2]

This paean to the horse as more than flesh mirrors the sagas of ancient Iceland, the Edda, where Sleipnir, the eight-legged horse of the god Odin, could traverse both land and sea and typifies the directions of the wind. In Norse legend Skinfaxi, whose name meant 'shining-mane', was the horse of the Day, while Hrimfaxi, 'frost-mane', was the horse of Night. These transcendent qualities travel from the days of myth to modern literature in J.R.R. Tolkien's heroic horse, Shadowfax, who elects to carry the wizard Gandalf in *The Lord of the Rings*. Tolkien was a great lover of the Edda and its influence is evident both in the great horse's name

This image, possibly a representation of the goddess Epona herself, has kept watch over the Vale of White Horse in southern Britain for more than 2,000 years.

and his qualities. He is swift and shining and 'Were the breath of
the west Wind to take a body visible, so would it appear.'[3]

In the culture of the ancient Celts the goddess Epona is closely
associated with horses, in particular through the huge figure,
almost 122 metres (400 feet) long, of a white horse carved into
the chalk hillside at Uffington on the Berkshire Downs (now part
of Oxfordshire). This image is thought to date from around the
first century BC and to be either linked with her cult or a zoomor-
phic representation of Epona herself. Epona's influence spread
to the occupying Roman army, who adopted her to protect their
horses and then carried her cult across Europe, usually depicting
her in a more lady-like form, riding side-saddle. Her origins are
prehistoric and she is linked across Celtic traditions to fertility
rituals and goddesses, including the Welsh Rhiannon, Irish
Macha of Ulster and Medb, or Maeve, of Connacht. These horse-
goddess figures link the understanding of the horse as a source
of spiritual energy with the female earth-mother figures and
emerge in the Celtic Christian tradition in the belief that the spiritual path is undertaken to unite all creation in harmony.

Another important aspect of Epona is that she not only appears as a horse and a woman upon a horse, but also as that strange animal, the horse–human hybrid. This creature has many guises and while it often emerges in traditions seeking balance between the physical and spiritual, such as Tibetan Buddhism, in the figure of the centaur of Greek mythology it seems conflicted. The centaurs were fathered by the monster Centaurus upon the mares of Pelion and were linked to the wild revelry associated with Dionysus, looking like satyrs in early images, but with the hind legs of a horse, rather than a goat. The most familiar image of them is with a man's torso and the body of a horse, and one idea of their derivation comes from the fame of the early cow-herders of Thessaly, who presumably were so skilled as to seem one with their horses, but were also famed for

A wood-engraving by Agnes Miller Parker in *The Fables of Esope*, 1931: because the stag encroaches on his territory, the horse allows the hunter to ride him. But the man refuses to release him and the horse's liberty is lost.

boorish coarseness, a feature common to centaurs. However, the other side of their nature may be seen in the character of Pholus, who entertained Hercules, and, more famously, Chiron, who was educated by Artemis and Appollo and took on the role of tutor to many heroes, including the young Achilles.

The wisdom of these centaurs makes a striking contrast to the reputation for savagery they held as a race, reflecting early ideas about conflicting human and animal nature, although the horse is generally depicted as a refined and intelligent creature in classical mythology. It is tempting to consider that it was perhaps not human nature that balanced out the animal passions of the horse in the hybrid figure, but the baser aspects of humanity that were refined in the union. Demeter goddess of women, marriage and agriculture, is sometimes depicted with a horse's head to recall her time disguised as a horse while she tried to avoid the advances of the god Poseidon. Poseidon, as mentioned above, was the father of Pegasus, and in his pursuit of Demeter assumed the form of a horse himself and fathered the horse Arion upon her. The acolytes who tended the temple of the goddess in her horse-headed aspect were known as 'Demeter's foals', which drew them under her matriarchal protection and, perhaps a dubious honour, into her dysfunctional family.

The love affairs and concerns of gods and goddesses that lead to horse–human hybrids are reflected in a more earthly context by traditions such as the Mari Lwyd of Wales. Particularly linked to the South Wales villages of Glamorgan, the Mari Lwyd appears during Christmas and New Year celebrations as the slightly alarming figure of a horse's skull, often decorated with ribbons, held on a pole by a man beneath a sheet, representing the horse's body. The head may sometimes be made of wood, but was originally a genuine skull. There is much debate over the meaning of the Mari Lwyd, with even the

Mari Lwyd belonging to the Llantrisant Folk Club in South Wales, who revived the tradition during the 1980s. It is made from a genuine horse's skull.

name being open to translation either as 'grey mare' or 'Holy Mary', but the figure, its name shortened and mutated in Welsh to *y Fari*, is carried from door to door around the villages. It may make its presence known by tapping on or peering through a window, and is accompanied by wassail-singers, sometimes with blackened faces and in their best clothes. As in many folk traditions, ritual verses are sung and this has developed into the *pwnco*, a rhyming contest between a member of the group and a householder, over permission to enter the house for a supper of cakes and ale. Once feasted, the Mari Lwyd group sing a farewell song and move on to repeat the performance, so, not surprisingly, the ritual is associated with revelry of the drunken kind, along with tricks and practical jokes. The wildness of this behaviour led to the suppression of the ritual during the rise of Methodism in Wales, so that it came close to dying out. Variations on the Mari Lwyd ritual include the demand for *y Fari* to dance a jig, and for a coachman to drive the figure along its route and eventually beat it to death. These more disturbing elements of the ritual suggest a link with death as well as the fertility and good luck generally considered to surround the tradition. Sometimes considered to relate to the flight of Mary and Joseph into Egypt with the infant Jesus, the roots of the Mari Lwyd are most likely wound around those of other rites of passage, in this instance the journey from the old year to the new. The ritual contains all the elements of eating, drinking and trickery associated with the symbolic crossing from one state to the next. This act of crossing the boundary reaffirms and acknowledges its existence, while the need for *y Fari* to be invited inside the house to perform the crossing confirms its provision of security for those inside.

The continuance of any custom over a long period of time inevitably results in changes of meaning and interpretation to incorporate or exclude ideas. Whatever the current or original meanings of this horse-headed figure, it belongs to a very old tradition. Palaeolithic drawings in Pin-hole Cave, Derbyshire, show a man wearing a horse-mask, and the worship of animals is thought to be one of the earliest forms of religious ritual. A link with Epona and similar early figures seems likely, especially as this tradition appears elsewhere also. *Y Fari* sometimes has a champing jaw, like its Kentish equivalent, the Hooden Horse, which is smaller and covered in a dark horse blanket so that it appears to be standing on all fours, unlike the tall stark white figure of Wales. Rituals such as the Mari Lwyd form a bridge between tales of heroic transcendence and quests for the spirit path, and the lives of ordinary people. By a paradox, the mystical becomes more accessible to everyday experience when it is explained away by magic and rooted in the earthy ground of the folktale.

This may be seen in particular in fairytales, where the two elements of the transcendent and the everyday are drawn together, and horses typify this by falling largely into two categories. There are those with more-than-mortal beauty, often ridden by heroic fairies or local supernatural figures with similar qualities, and much more ordinary horses, often notable for being plain or shaggy coated, but with particular qualities of intelligence.

In her study of Irish legends in 1887, Lady Francesca Speranza Wilde, mother of Oscar Wilde, declares the horses of the *Daoine Sidhe*, the most aristocratic of the heroic fairies, to be as splendid as their riders:

> And the breed of horses they reared could not be surpassed in the world – fleet as the wind, with the arched neck and the broad chest and the quivering nostril, and the large eye that showed they were made of fire and flame, and not of dull heavy earth.'[4]

As these heroic fairies are usually of human or more than human size, and have links with the highest ideals of the medieval romance, their horses, unsurprisingly, match them. In Lady Wilde's description they sound very much like Spanish or Arab-type horses, which were first imported into the British Isles in the medieval period, rather than the small and hairy native ponies. It is possible to see the conflicting logic of aspiration and practicality in these tales. The fairy horses have all the grace and rarefied beauty of breeds foreign to the cold damp air of Britain, but the native breeds, short-legged and shaggy, are better suited for the reality of life. When this second type appears in traditional fairy stories, they often do so as creatures whose worth is initially undervalued but then becomes clear with time and the emotional growth of the hero.

In the traditional Celtic tale 'The Young King of Easiadh Ruadh', the hero wins a lovely queen to be his wife when he beats a Gruagach, an Irish trickster giant, at chess. With hubris as traditional as his bravery and courage, the young king cannot resist trying his luck against the giant again. To help him, his wife tells him that although he will not be impressed by the next prize, a dun-coloured shaggy-coated filly, he should accept and value her. This shaggy filly becomes both his rescuer and teacher when he insists on playing chess again, and through the intervention of the horse, the woman and other companion animals, the young king grows, as all heroes do, by being humbled and gaining wisdom.

Shaggy horses turn up in many fairy tales and legends of colder climates, and often echo this pattern of the horse that is wise, rather than beautiful, and acts as advisor to the adventurer/ hero. Often the animal is killed, or even asks to be sacrificed for the sake of the human companion to whom it has devoted its life. The talking horse, sleek or shaggy, rarely speaks for its own sake, and is sometimes given the gift of speech so that it may unburden itself of its long grief over human cruelty and be fully understood.

M. Oldfield Howey, in his comprehensive study of the horse in myth and legend, first published in 1923, takes an interesting stance on tales of talking horses. He relates that 'Indian myth generally seems to speak of the horse as a fully developed self-conscious creature with powers (e.g. of speech) which it certainly does not now normally possess, existing long ante rior to the creation of man'.[5] Howey leaves open the possibility that horses once had the power of speech and maybe still would have, had humans only the ability to hear what they were saying.

The people of Turkic Siberia are rooted in the very ancient Asian traditions of ancestors from Tuva and Khassia, and the hero's horse is a central figure in folk tales from this source. The hero's ability to hear the voice of the horse is crucial to his growth and survival. The hero must tame and be tamed by the horse, which will then devote itself to the hero, acting as an adviser and as a balance to the hero's tendency towards recklessness. In this tradition, as in others, the horse embodies the life-force, acting as a bridge between the inner and outer worlds, or the everyday and the spirit world. Siberian shamans consider their ritual drum to be a spiritual horse upon which they ride across the boundary between the worlds. The shaman's cloak incorporates materials such as hair, feather and bone to become part animal, part bird, often with a fringe to symbolize a horse's tail on the back. The Siberian storyteller's musical instrument, the *igil*, has strings of horsehair and is used in the distinctive style of singing known as 'overtone' or 'throat singing', where the singer creates two or more pitches at the same time. The origin of this type of singing and playing is linked to calling the spirits. In one story a horse teaches his human friend how to create the instrument, but the man must learn to play, illustrating once more the interdependent nature of the relationship between the hero and his horse.

Supernatural creatures disguised as horses turn this sort of tale on its head when goblins see the mischief potential of human expectations and dependence on the horse as companion and beast of burden. Common in British folk tales, supernatural horses usually appear in traveller's legends and myths surrounding difficult terrain, such as moorland, being

Slide lantern image of a Goldes Shaman of Siberia, c. 1895.

linked to the trials of the journey. These heartless goblins, disguised as horses, lure the traveller with the promise of a ride home on a cold night and then disappear, or deposit him in a ditch and are heard to run away neighing with laughter. The brag of England's northern counties, the shag-foal or tatter-foal of the Lincolnshire Fens, the Picker-tree Brag and the Dunnie of Northumbria are all disguised goblins that behave in a similar tricksy manner, laughing at the distress of their hapless riders, who end up in rivers and ditches, having lost their goods or passengers, and are left to make a wet walk home in the dark.

Christopher Marlowe seems to draw on these familiar tales in *Dr Faustus* (*c.* 1590), when Faustus, having sold his soul to the Devil in return for magic powers, conjures a horse out of hay to trick a horse-courser, or dealer. He warns the horse-courser: 'Now, sirrah, I must tell you that you may ride him o'er hedge and ditch and spare him not; but, do you hear? In any case ride him not into the water.'[6] Although running water has a long association with the breaking of spells, human greed outweighs any suspicions the horse-courser might have and he rides the hay-horse into a stream straight away, hoping to uncover some magical potential to make it worth more than he paid for it. He is plunged into the water at once, as the magic dissolves and the horse along with it. The humour in this would have an impact upon a late sixteenth-century audience, familiar with the dangers of magic, that is lost to us today. As Faustus is tricking a con-man to make a point, the question arises as to whether this is at the heart of these legends, the equivalent of the fisherman's tale of the 'one that got away'. As few riders will admit but most will know, owning up to 'falling off' is rare. Most accounts of parting company with a horse tend to tell of being 'thrown' or 'bucked off', rather than a simple admission of a loss of balance or concentration at the wrong moment. Maybe travellers on the old roads of Britain, trekking across boggy marshland under a pale moon, were reluctant to admit that they nodded off and fell from their horse, who disappeared, alarmed, into the mists. No doubt it made a better tale once the horse became a disguised goblin, out to trick them.

Horses tend also to have what might be classed as 'urban legends' attached to them, based on ways of deciding their character and worth. Beliefs based on the number of white legs a horse should have appeared in print as long ago as the sixteenth century and as recently as 1958, although they often contradict one another. In 1560 Sir Thomas Blundeville, loosely translating the work of Italian riding master Federigo Grisone, devotes five pages to a discussion on the amount of white that can safely be overlooked on a horse's legs, too much, of course, being 'an evil signe, betokening debility'.[7] In 1958 advice to those in search of a good horse in America included that a horse with one white sock is always the strongest, that a white leg is the sign of a weak horse and that one should always buy a horse with three white stockings. It seems that the amount of white makes a significant difference, as a sock, a stocking and a leg, measuring different extents of white on any one limb, indicate. The most interesting feature about ideas like this, which any experience will show to have no basis at all, is the way in which they turn up in widely separated cultures.

A traditional Arabian folktale tells of a man who was expecting the first foal from his beautiful mare, with all his friends gathered to witness the birth and bid to buy the foal as soon as it was born. The first sight of the foal's head showed a star: the owner was delighted, because this was a sign of great good fortune. Then a solid-coloured near front leg appeared and the observers started bidding high prices. But then the other front leg appeared with a white sock. The bids were cut by half. The arrival of a white near-hind leg made the owner rejoice – this foal was not for sale for any money! But then another white hind leg appeared and the owner wept, ordering that the worthless foal be destroyed immediately.

Something of a tragedy for the mare and her foal, this story may well suggest, like the disappearing goblin-horses, ways for humans to explain their own mistakes and preferences, which over time become assimilated into beliefs. William Cavendish offers some common-sense advice, believing that

> endeavouring to discover the constitution and particular disposition of horses by their marks and colours . . . seems a kind of conjuration or sorcery to me. For should

This horse has the blue-grey hoof long associated with durability and soundness. His 'shoo-fly' throat-tassel and binding to protect his long tail-hairs suggest that he is highly valued.

Tab. x

Monoceros Unicornu.
Einhorn.

Capricornu Marinu
Meer Steinbock

Monoceros Unicornu
Einhorn

Mythical creatures appear in many early natural history books, such as these unicorns from Jan Johnston's *Beschrijving Van de Natuur der Viervoetige Dieren*, 1660.

these marks happen to be accidentally true, the cause does not proceed from the colour of the foot, as from the quantity of spirits in the horse's nature.'[8]

Yet for all this obvious wisdom, even the modern would-be horse owner is likely to have heard:

One white foot, buy a horse;
Two white feet, try a horse;
Three white feet, look well about him;
Four white feet, do without him.

While this rhyme is unlikely to have much attention paid to it today, that it is nevertheless well known suggests that superstitions have a more lasting impact than is readily admitted. It is still common to hear that horses with a blue or wall eye are bad tempered and that white hooves are weak, while everyone knows that a chestnut mare, like a red-headed woman, can be fiery.

The way in which modern myths about horses and horse-keeping find their roots in some of the oldest traditions demonstrates the influence of the horse upon human thinking at an instinctive level. Covering so wide a breadth of the world, through folk-tales, religions and mythology, it not surprising that many of the tales cross over or are found in several forms with no obvious point of contact. The imagery of the horse has a resonance that crosses cultural lines to become a shared discourse. As Howard Schwartz says, 'the boundaries of folklore are no more defined than those of the waves of the sea',[9] so when we can travel the world on horseback, it is perhaps inevitable that tales of the horse will travel equally far.

3 The Man-made Horse

One of the most remarkable features in our domesticated
races is that we see in them adaptation, not indeed to the
animal's or plant's own good, but to man's use or fancy.
Charles Darwin, *The Origin of Species*

The desire of humans to shape the world to our use has impacted strongly on the development of the horse. Indigenous breeds,
or those which have become seen as indigenous, are shaped by
their environment and it is usually a particular feature of natural
quality that makes a horse suitable for a specific activity or work.
By controlled breeding, which goes back 4,000 years, humans
have been able to produce horses that are more suited to pulling
or speed, more agile or athletic, taller or shorter, and to at least
aim for certain colours or markings, though nature tends to out-
wit the most sophisticated plans in this respect.

The Shetland Pony, named for its remote island home 160
km (100 miles) off the north coast of Scotland, has short legs, a
thick coat and the robust ability to thrive on the most sparse of
diets. These are the essential features that have shaped the
breed and enabled its survival on the windswept islands of its
native home. It is thought to have been introduced to the
Shetland Islands, possibly from Scandinavia, as long ago as the
Bronze Age, while the 'Monk's Stone' from the island of Burra,
which shows a monk riding a small pony, dates from AD *c*. 800.
The standard measurement for horses is the 4-inch 'hand',
which roughly relates to the width of a man's hand, with a pony
being under fifteen hands high. The Shetland Pony, however, is
traditionally measured in inches and averages around 40 at the

wither, the point of the shoulder. The Court Books of Shetland refer to its role as a working pony in the early seventeenth century and specific mention of its hardiness in the challenging environment is made in 1730.

The value of these small creatures for ploughing, carrying seaweed and peat and general heavy work out of all proportion to their diminutive size was appreciated by the Shetland Islanders long before the 1840s, when a ban on women and children working in coal mines led to a new role away from home as pit-ponies. Since then, Shetland ponies have become popular for children and as driving ponies, as they combine small size with

A study of coat patterns in horses of the Imperial Stables at Versailles, by Théodore Géricault, c. 1813–14. The differences in muscle development suggest that each horse has a specific working role.

character and intelligence. In all these roles the natural features of the Shetland Pony have been the foundation of its uses. But the Shetland is also among the many breeds whose natural characteristics have been developed to suit human purpose. Having left their native home, they were exported widely and what is today known as the American Shetland would be barely recognized as family by its hairy Scottish cousins. It has been cross-bred with other breeds and types to develop a much finer animal, high-stepping and dynamic, though with the aim of retaining the hardiness of its ancestors. There are also Miniature Shetlands, which may be smaller than a Great Dane dog, but are very similar to the full-size ponies in appearance and fully equipped with character. The Shetland Pony's story is simply one of many, and breeds have been altered, refined, 'improved', persecuted, nurtured and even wiped out completely, depending on the role they are seen as playing in relation to humans.

Elwyn Hartley-Edwards, discussing native British ponies whose names are now only a romantic memory, such as the Goonhilly, the Lincolnshire Fen pony and the Irish Hobby, says,

The Miniature Shetland – small size, big personality.

THE OLD ENGLISH BLACK HORSE.
Stallion, in old Blacklegs from a Mare of the Thicklegs breed... bred in W. Drummonds at Houghton Heads.

A lost breed – the Old English Black Horse by William Nicholson, based on a 19th-century painting by William Shiels.

'These ponies became extinct either because there was no further use for them in a changing society or because they were absorbed by more fashionable strains.'[1] This is a telling comment, and while it is true that humans have regarded horses as a resource to be shaped or disposed of at will, it is also true that a great deal of dedication has gone into preserving, promoting and saving distinct horse breeds. Horse breeding today is a highly competitive and expensive field but also, very often, a labour of love.

Every breed of horse has its particular story, but a few selections from across the world offer insights into a broader picture. The English Thoroughbred and the Appaloosa have been influenced very strongly by humans, while the Icelandic Horse is a prime example of what happens when humans leave well alone. The Arab, however, is a key breed not only for its natural qualities, but also for its influence on other breeds due to prepotency, the power of passing on its hereditary features of intelligence, stamina and beauty to any breed with which it is crossed.

The Arab horse has a long history. Rock paintings and inscriptions in southern Turkey dated to 8,000 BC show a horse with the familiar Arab characteristics: a dished face, refined head and delicate strength, along with the high-tail carriage. Images that illustrate the consistency of the breed through these features may be found in ancient art work along the Arabian Peninsula and records from early travellers. Marco Polo comments on the export of horses from the Yemen to India in the thirteenth century and horses from the coastal plains of Oman are mentioned in traders' records from the early sixteenth century. They are also known to have been exported in the thirteenth and fourteenth centuries.

Arab horsemen by Adolf Schreyer (1828–1899).

Debate over the original location of the earliest Arab-type horses is ongoing, but it is generally accepted that they came from the northern edges of the Fertile Crescent or the south-western corner of the Arabian Peninsula. In both these areas, natural pastures would have once existed to nurture the development of early types. The Arab is, however, very much a breed that has lived alongside and been shaped by humans, since they were first domesticated by the Bedouin nomads of the Arabian Peninsula shortly after they adopted the use of the camel. The camel's ability to both go without and to carry large amounts of water was crucial to life in the desert for both humans and their horses, as a horse drinks around eight times as much as a camel. Camel milk has also been a traditional food for Arab horses throughout the centuries until the present. Sheikh Zayed Bin-Sultan Al-Nahyan (1918–2004), the Founder and President of the United Arab Emirates, a great supporter of the Arab horse, kept a herd of milking camels for his mares and described the camel as 'the companion and support of the Arabian horse'.[2]

The demands of desert life bred endurance into people and horses and fostered a very close relationship. Horses had to thrive on sparse feeding and low water rations in harsh dry conditions, while being ready to race at speed into the fierce skirmishes of desert warfare among the different tribes. The bloodlines were preserved by oral tradition for centuries and the first written pedigrees of the fourteenth century were based on these traditions, which were preserved as carefully as human family trees and always traced through the female line. The *Asil* horses, those of the purest bloodlines, were never bred to non-*Asil* horses and poor animals were not allowed to breed at all so, over time, a foundation stock of unfailing quality was established.

The Arab was primarily a warhorse, with camels being used as beasts of burden, and, in contrast to Western traditions,

mares were ridden into battle. Chosen over stallions for being
less likely to cry out in challenge to other horses, the war-
mares were so treasured that they were often stabled within the
family tent and kept with the children. Thus their disposition
too became prized and their contact with humans developed
sensitivity and gentleness, while selective breeding enhanced

their great beauty. Over time, these traditional breeding practices led to a horse honed to its natural characteristics in their finest form, made hardy by its desert home with genetic purity giving the natural tendency to pass its finest qualities on to its progeny. The Arab became the progenitor of breeds as different as the Australian Waler, the Orlov Trotter and, perhaps most famously, the English Thoroughbred.

It seems perhaps a little strange that a horse founded on desert stock should be described either as 'English' or 'Thoroughbred', but in the intervening years the breed has developed into something distinct from its forefathers. Today's

Thoroughbred is known primarily as a racehorse, though racing in England already had a long tradition, rising to high popularity in the reign of Charles II (1660–1685), which predated the
establishment of the breed. Studs for 'running horses' were well
established by the seventeenth century but these would have
been crossbreds rather than any specific breed, with the blood
of native and imported stock.

The first date of the arrival of a purebred Arab horse in
Britain is debated, although it may have been as early as the
tenth century. Sir Thomas Blundeville refers to 'Turkey Horses'
in his horsemanship manual of 1550 but at that time much of
Arabia and Syria was part of the Turkish Empire, so horses from
several Eastern locations could have been embraced by that
term.[3] However, it was not until 1580 that a Turkish ambassador
was sent to the court of Queen Elizabeth I to open trade negoti-
ations, so it seems unlikely that many quality Arab horses
arrived prior to this time. In 1667 William Cavendish, by then

Duke of Newcastle and highly acclaimed, not least by himself, for his knowledge of horses, wrote of Arabs, 'I never saw any but one of these Horses', which he remembers was sold for £500 by a Mr Markham to King James I, suggesting a date between 1603 and 1625. A horse known as the Markham Arabian is mentioned in the 1891 edition of the Stud Book as having been said to be the first of its kind in England, though the entry casts doubt on the truth of that tradition. This seems likely to have been the same horse, and it is clear that, whether the very first or not, it must have been one of very few at the time to have warranted particular mention by different authors. Newcastle reports rumours that a 'Right Arabian horse' could fetch from £1,000 to £3,000, which he considers an 'Intollerable and Incredible Price'.[4] As a good Arab youngster could be found for that price today, he may well have been right, and he would be startled by the hundreds of thousands paid now for the very finest examples of the breed.

By the 1730s horses of desert origin were less rare and the three sires that would found the Thoroughbred were in British hands. They were the Darley Arabian, the Godolphin Arabian and the Byerly Turk, all horses named for their owners, as was the custom at the time. Of these only the Darley Arabian was certified to have been of purebred Arabian blood, though the tradition that they were all Arab horses remains strong, if frequently disputed. But even if two of them were Turkish, the influence of the Arab horse would have shaped their development. From these three stallions and thirty 'tap-root' mares, the 'running horse' of the late seventeenth century developed into the English Thoroughbred, a designation first used in 1821, but known world-wide today.

The Thoroughbred is a prime example of a horse developed by man for one particular purpose, namely to run at great speed. In this aim the Arab blood was not intended primarily to provide the swiftness but rather the potential to remain true to

Mambrino by George Stubbs, c. 1790. An important stallion in the development of both the English Thoroughbred and American Standardbred, Mambrino was directly descended from the Darley Arabian.

type, essential when founding a breed. None of the founding stallions ever won races, but 81 per cent of modern Thoroughbred genes derive from 31 named ancestors, who in turn all descend through the male line from one of these three desert stallions.

In *The Origin of Species* Charles Darwin repeatedly refers to the development of the racehorse as an example of the way in which man can use natural selection to his own ends. When considering the process of domestication, he says:

> Some effect may be attributed to the direct and definite action of the external conditions of life, and some to habit; but he would be a bold man who would account by

"NUTWITH," THE WINNER OF THE ST. LEGER DRAWN BY J. F. HERRING, ESQ.— C 3431 page.

such agencies for the differences between a dray- and race- horse . . . The key is man's power of accumulative selection: nature gives successive variations; man adds them up in certain directions useful to him.

He notes that it is this process that has led to the English race-horse being taller and faster than its parent Arab stock and goes on to refer specifically to one of the first great racehorses to illustrate his point:

For instance, there must be a limit to the fleetness of any terrestrial animal, as this will be determined by the friction to be overcome, the weight of body to be carried, and the power of contraction in the muscular fibres. But what concerns us is that the domestic varieties of the same species differ from each other in almost every character, which man has attended to and selected, more than do the distinct species of the same genera . . . With respect to fleetness, which depends on many bodily characters, Eclipse was far fleeter, and a dray horse is incomparably stronger than any two natural species belonging to the same genus.[5]

Eclipse, foaled in 1764, was never beaten on the track and won eighteen races between 1769 and 1770 with apparently little effort. A great-great-grandson of the Darley Arabian, he was only one of several prodigious stallions that arose from the three bloodlines. The most famous included Flying Childers, foaled in 1714, son of the Darley Arabian and said to be the fastest horse ever born, Herod, a grandson of the Byerly Turk, foaled in 1758, whose progeny won over 1,000 races, and Cade, a 1734 son of the Godolphin Arabian, whose offspring through the female line founded the great harness racing breed, the American Standardbred. Other stallions who are less well known today nevertheless also contributed Arabian blood to the race track, as their names show, including the Unknown Arabian, D'Arcy's Chestnut Arabian, the Leedes Arabian and Alcock's Arabian. Alcock's Arabian is considered to be one of the two horses responsible for introducing the colour grey to the Thoroughbred line. From 1770 Arab blood was no longer introduced as the breed was considered to be established and home-bred horses were used to continue the blood lines.

Since then the Thoroughbred has become the centre of a thriving and expensive international sport. The race-track at Newmarket, founded by Charles II, remains one of the main racing centres in the world today, largely due to the way in which the Thoroughbred has been shaped for its task. It is a horse of graceful proportions, clean-limbed and powerful in the quarters and with long sloping withers to encourage an

economical range of motion for forward movement, unlike breeds with elaborate action and high-stepping strides, which would be unsuited for the race track. The Thoroughbred is the epitome of the man-made horse, and while this makes it an animal which is treated with the greatest of care when it is successful, poor examples and old or failed racehorses can have difficult lives.

Viva Pataca, winner of the Hong Kong Champions & Chater Cup in 2006 and 2007.

The story of the Appaloosa is very different and it is only through dedication that the breed survived into the twentieth century. Today it is known not only for its spectacular spotted coat colours but also for its part in the history of the Nez Perce people of the north-east corner of Oregon. The spotted gene has deep roots, as discussed in chapter One, and the coat patterns of spots or splashes of dark colour may be a relic of the need for camouflage. The spotted horses of America are thought to have derived from spotted strains in the Spanish horses imported by the conquistadors in the sixteenth century and when the Nez Perce of the Palouse Valley in Oregon took a liking to these colourful mounts, a decisive moment in the history of both horses and humans occurred.

The Nez Perce were a people who readily adapted and developed new ideas from outsiders and were the first Native Americans to take on the practice of selective breeding. A settled tribe of gatherers who relied on the salmon run rather than the wandering buffalo for food, they used their spotted horses primarily for racing and ceremony. In time the horses they bred became known as 'Appaloosas', from 'a Palouse horse' and had several striking colourations. A leopard-spot has a light-coloured coat and either chestnut or black markings all over the body. A blanket-spot has a solid-coloured body with a white rump marked with dark spots, while a snowflake has a dark body with white spots. There are many variations on these patterns and while some horses remain much the colour they are born throughout their lives, others

The mottled muzzle and glimpse of a white sclera define this Appaloosa as clearly as his spots.

change considerably over the years. Another distinctive feature of the Appaloosa is the sclera, a white ring around the eye, which can give a rather worried expression, though generally they have a good temperament, friendly nature and the ability to learn readily.

They appealed to the Nez Perce as colourful, hardy and practical so by the mid-1700s their herds were well established and are mentioned in the 1806 records of the Lewis and Clarke expedition. The Nez Perce had helped the expedition and initially been proud of their friendship with the white settlers but, in 1855, Governor Isaac Stevens began to pressure them to give up their ancient tribal lands and stay in a place allotted to them by the newcomers. The idea that land should have clearly defined boundaries or owners was foreign to the native people; their belief that land could belong only to the Creator was foreign to Governor Stevens. Some chiefs signed lands away on the understanding that what they were allowed to retain would be safe for their people. But a new treaty in 1863 took away over three-quarters of the land they had left, including the Wallowa Valley, where those of the Nez Perce who had refused to sign at all had made their homes. The people of Wallowa Valley maintained a policy of peaceful resistance for fourteen years, even as they were increasingly threatened and pressurized towards moving.

At last, while the tribe were struggling to cross a river swollen by meltwater, a party of white men stole some of their cattle and, in the hurry to get the rest across the river before they could be stolen too, many were drowned. While their most famous leader, Heinmot Tooyalaket, more widely known as Chief Joseph, attempted to calm their anger and keep peaceful relations possible, a small band of warriors sought revenge and killed eleven white men during a night raid. Chief Joseph

Account of the flight of the Nez Perce, in *Harper's Weekly* (27 October 1877).

said he would gladly have given his own life to undo their actions but sided with his warriors, understanding their fury. As he attempted to gather his people and stock to leave without further bloodshed, the army attacked. Despite his dedication to peaceful means, Chief Joseph was an astute military leader and, though outnumbered by two to one, the Nez Perce routed their attackers and made their escape. With the whole tribe of 700 people on the move, of whom only 250 were warriors, carrying all their goods and herding 2,000 horses, the Nez Perce fled on a wild journey over rough terrain. They knew that they could never return to their own lands or even accept life on the reservation now without punishment, so they set out to cross the Canadian border, over 1,300 miles away.

Thinking they had reached their goal, they stopped, starving and exhausted, but had miscalculated and were just short of Canada, in the Bear Paw Mountains of Montana. There, in October 1877, the army launched a surprise attack. Almost 60 women and children were among the 80 Nez Perce killed before they were forced to surrender both themselves and their horses, which were all confiscated, many to be slaughtered. The spotted horses had become so great a symbol of resistance that from then on the Army was under written orders to destroy every one that it came across, and in one instance, 400 head were driven into a canyon and shot. As the Nez Perce accepted the life of farming they had been forced into, they were allowed to keep a few of their horses on the condition they were bred to draft stock, making the offspring unsuitable for fighting or hunting.

By 1937 there were only a few hundred Appaloosas left in the world, but one farsighted Oregon rancher, Claude Thompson, set out to gather breeding stock from those that had survived. Slowly a revival began and, in 1938, the Appaloosa Horse Club was formed in Moscow, Idaho. Since then the Appaloosa has become one of the largest breed registries in the world and these versatile horses have a secure future, being popular in many disciplines from trail-riding to show-jumping.

In this instance humans were responsible for the rise, fall and recovery of the Appaloosa, reflecting the way in which horses are at the mercy of humans. The epic journey of Chief Joseph and his tribe is recreated by the Appaloosa Horse Club as an act of memorial and each year a section of the route is ridden by up to 200 riders, all on registered Appaloosa horses. The philosophies of Chief Joseph and his dedication to the survival of his

Hide painting depicting a horse-stealing raid. The reverse colouring of white on black in the spotted horses would be known as a 'snowflake' in a modern Appaloosa.

people have become synonymous with a desire for peaceful co-existence between humans. When he died in 1904, the cause of death was recorded as a broken heart.

Whereas wide dissemination of their bloodlines has been the key to the survival of several breeds, for others it is isolation that has ensured their survival. One of the most notable of these is the Icelandic Horse, preserved through a deliberate policy of isolation, which became law in 1882. Even today, an Icelandic Horse that leaves Iceland cannot be returned. Iceland originally had no indigenous horses, but the breed whose development is so closely linked to the terrain and isolation of Iceland first arrived there from Norway with settlers in the ninth century, to be followed by native animals with later settlers from Norse colonies in Scandinavia, the Western Isles of Scotland and Ireland. The foundation stock from which the Icelandic horse developed was drawn from the sturdy types of these regions. It

Shaped by its environment – the Icelandic Horse.

is widely believed that to keep the horse population strong and disease free, the Icelandic Parliament decreed in the tenth century that no other horses could enter Iceland. There is debate over the factual basis for this tradition and it may simply have been the difficulty of transporting horses to Iceland that kept the breed pure. However, the law passed in 1882 remains in place today and has ensured that the Icelandic is one of the few truly pure breeds, with direct descent from that early stock. This has ensured that the majority of contagious equine diseases are unknown in Iceland and the breed has retained all its native characteristics. A policy of culling and eating weak stock also ensures that only the very strongest horses survive to pass on their genes. There is no Icelandic word for 'pony' and, although Icelandics rarely stand much above fourteen hands high, they are always known as horses and have bone density which gives them the carrying capacity of a much larger animal, being fully able to carry a grown man. Slow to mature, they are traditionally left to run wild in a herd until they are at least four years old, but they are long lived and the oldest Icelandic Horse on record reached around 55 years of age.

Icelandic Horses come in a wide range of colours, with some which are unusual, such as a blue dun, or a silver dapple, and they have developed unusual gaits too, to cope with the difficult terrain. Alongside the familiar walk, trot, canter and gallop, Icelandics have a four-beat lateral gait called the *tölt*, a ground-covering pace which is very comfortable for a rider. Some also have a fifth gait, the flying pace, a two-beat lateral gait that can reach speeds of 48 km (30 miles) per hour and is often used for racing. These gaits are natural to the breed and even foals display them while playing. These horses are considered very steady and forward-going, meaning that they are less inclined to take fright or be reluctant than some breeds. They are also known for making strong bonds with a rider they trust and their surefootedness has made them invaluable in the terrain of their home and popular in other countries as well. They are seen today in all sorts of equestrian sports and activities from endurance riding to children's Pony Club. Their excellent disposition makes them easy to handle while their small size and great strength means that they can carry an adult with ease while being economical to keep.

The native horse as representative of its homeland on an Icelandic stamp.

The practical value of the Icelandic Horse in its homeland has continued to the present day in accessing the country's wild landscapes in a way which does not damage the eco-system. Up until a few decades ago, major roads were still few in Iceland so the horse had a vital role to play as a pack-animal and a means of transport. The traditional method of riding one horse while back-up mounts run free until they are needed continues today and the homing instinct of these horses is so reliable that it is not unusual to borrow a horse and then simply release it, knowing it will find its own way home. Today, riding holidays on Icelandic horses are a popular tourist attraction but any rider taking their own hat or boots to ride in has to disinfect them before they can enter the country and any used imported tack or gear has to go through rigorous disinfecting procedures.

The environment of Iceland demands toughness and resilience from any creature that lives there. Harsh weather, volcanic eruptions and sparse feeding have all honed the breed to exceptional character and toughness. At the height of their use as a working animal, Icelandic horses pulled carts and carried packs and had to be able to swim glacial rivers carrying a rider or goods. They carried midwives to births and the dead to burial, often ridden only in a sheepskin saddle with a loop of rope around the lower jaw for a bridle, a style also used by Native American riders. Unlike many breeds, it was the elements, not humans, that shaped the Icelandic Horse over time. Wild, cold and windy weather encouraged them to be small, hardy and long-coated for warmth. Feathered legs helped to shed water and a strong digestive system and efficient metabolism means that they can survive on the meanest of diets and are liable to run to fat on good grazing. Rough terrain helped strong hooves and sure-footedness to develop while the difficult environment also encouraged resilience and endurance. The reliance of horse and rider upon one another is heightened in any difficult situation and value placed upon these horses by Icelanders is an acknowledgement of their importance to the development of the people. It is traditional to give all horses of the breed Icelandic names, even if they were born far away and will never see the land of their ancestors.

Human perceptions of the horse as a resource have directly resulted in all the breeds and types of horses, even those seen as wild, that are spread across the world today. The way in which those breeds and types have been developed has often been ruthless or even covert. When Henry VII banned the export of any horse worth more than 6s 8d from Britain in 1495, he simply encouraged horse smuggling, so that by 1531 Henry VIII had made the unlicensed export of horses a felony. This included the export of horses into Scotland, for fear that they would strengthen the Scottish military and return ridden by an attacking force. Still unsuccessful, the ban was re-enacted a second time by Elizabeth I in 1562, even though by then the punishment could be up to a year's imprisonment. A steady stream of horses nevertheless left England over land and sea, while they were also imported in their thousands, particularly during the reign of Henry VIII.

The movement of horses around the world for reasons of trade, practicality and, very often, the status symbol they pro-

England's King John hunting deer, c. 1210. The development of horses for different noble pursuits is reflected in these two images of the hunting field and the tournament.

Mort parole intreus in conte al.
Si mid colp lin dona droite esple poignal.

Li eschu litragi cose fust de cendal.
For fu lobrs to paues criminal.

vide, has impacted on the development and even the survival of the different breeds. Maintaining the purity of bloodlines, as with Arab and Icelandic horses, or the careful management of cross-breeding, as in the development of the English Thoroughbred, has made the modern horse arguably as much the product of intellect and science as of nature. While this has saved several breeds from extinction, it has resulted in others being lost altogether but, as Darwin observed, the demands of human purpose create their own agenda. The ethics of considering any other living thing as raw material are under more debate today than perhaps ever before and the future relationship between humans and horses is likely to shift with the ideas of changing times. The fact that there are so few truly wild or native breeds left places a huge responsibility upon human shoulders and it seems that the future of the equine gene pool is completely at the mercy of human wisdom and whim.

4 Riding into History

The first pact man ever made was with fire. The second pact
was with the horse. With fire he mastered his environment,
but with the horse he conquered the world.
Hank Wangford, *Lost Cowboys*

The horse has enabled many aspects of human development,
shaping physical and cultural landscapes in what is usually seen
as a supporting role. In China horses were essential across a vast
and challenging landscape both in war and the day-to-day
management of transport, communication and supplies.
Horses and chariots were entombed with their owners from as
early as 1600 BC to accompany them into the afterlife and during the Western Zhou Dynasty, around 1100–771 BC, military
might was measured in available chariots.

The introduction of new equine bloodlines to Britain during
the Roman occupation introduced hybrid vigour along with
greater size and strength to the small native stock. The value
placed on horses in Wales is evident from as early as AD 950,
when riding a horse with a saddle that galled its back could incur
a fine of four pence, a figure that would be quadrupled if the skin
was broken. During the medieval period in Wales a growing avail-
ability of horses transformed travel, trade and battle tactics. In
the eleventh century Robert de Belêsme imported horses from
Spain to Powys and established a tradition of fine horse stock so
successful that by the Battle of Falkirk in Scotland in 1298 one
third of the horses in Edward I's army had been bred there.

While a similar role has been played by the horse in the
establishment of civilizations across the world, there are also

Japanese 'Haniwa' earthenware horse. Figures of this type were used to decorate and guard tombs from 29 BC to the sixth century AD and replaced earlier sacrifices of living animals and servants belonging to the deceased.

those cultures for which the horse became a feature of personal and cultural identity so intrinsic that the relationship, changing and mutating over time, has survived to the present day. One of the best known horseback cultures is that of Mongolia, which was already well established by the time the thirteenth-century warriors of Genghis, or Chinggis, Khan rode into history. Genghis Khan's complex and savage vision began with the forcible union of his people to make a stand against the gradual ascendancy of China. With his own territory secured, he set out a campaign of expansion into Asia Minor, Russia, Persia, Europe and India that went beyond his own lifetime and was continued by his successors over the next century.

The vast numbers of Genghis Khan's cavalry were the key to his success and, to create his Black Standard, every unit gave hair from each of its horses to be literally woven together as a powerful symbol of massed strength. The small and hardy native horses were honed by generations of survival on the steppes, an exposed landscape of plentiful grazing but severe weather. Genghis Khan's armies travelled up to 130 km (80 miles) a day

This finely caparisoned ceramic funerary horse is glazed in pale green and dates to the Sui dynasty of China. AD 581–618.

with huge herds of horses and he established laws that ensured a vast available supply for all purposes, from cavalry mounts to pack animals. Among these laws were routine gelding of all but the best males to help manage the horses while refining breeding stock and enabling them to be kept in one group without fighting. Mares were limited to the number one stallion could cope with both in terms of fertility and control, while young stock went through a routine of early handling and riding to ensure that they were tractable by the time they came to maturity. A strict programme involving complete rest after a hard campaign, followed by gradual reintroduction to work, underpinned the success of an army that was primarily cavalry.

The skill of the riders was developed during a childhood in the saddle so that the hard gallop of the battlefield became second nature. Using terror to subdue his enemies and his own subjects, Genghis Khan maintained his position through a fierce code of justice that meant that even petty crimes could result in instant beheading. But good service to this uncompromising

warlord resulted in his loyal care and under his rule there was a high level of law and order. Art, culture and trade flourished, while it was said that a virgin carrying a pot of gold could walk the length of the land safely. Yet safety remained relative, and in 1227, when Genghis Khan died, 1,000 horses were ridden across his grave so that the trampling of their hooves would hide its whereabouts. Horses were the key to his success and to his peaceful resting place. Their riders, however, were slaughtered so that its location remained a secret.

Genghis Khan's cavalry: a 16th-century illustration of the 14th-century Persian account of *The History of the Mongols*.

The horseback culture of the nomadic people of Mongolia has never faltered and today the Festival of Naadam, dating back to Genghis Khan's training grounds, remains central in their calendar. Held in July, it celebrates the 'manly games' of horse racing, archery and wrestling, though women can enter all events except the wrestling. Alongside the games the festival has

Young Mongolian riders prepare for the Naadam Festival Horse Race; note the spotted pony ridden by the adult.

parades, music and military displays in traditional uniforms. It is held as an annual ritual to honour the mountain gods, but also to celebrate community endeavour. The climax of Naadam is a horse race where the riders are between five and thirteen years old. The five winning horses are immortalized in poetry while the youngest losing horse is rewarded with a song for its efforts.

This festival illustrates the continuing traditions of a nomadic culture in the modern world that influences its ancient but changing society. Today Mongolia is a popular tourist destination, leading to debate over the impact and benefits of cultural tourism in view of the need for a sustainable future for the people of this wild landscape and the horses still central to their lives. The President of the Mongolian Association for Conservation of Nature and the Environment, J. Tserendeleg, says: 'It is not possible to imagine Mongolian history without horses, and I think it is not possible to view the future of Mongolia without horses as well. Mongolia is not Mongolia without horses.'[1]

By way of complete contrast, while Genghis Khan's army was galloping across the Mongolian heartlands, the export of the desert horse from its native lands was laying the groundwork for an equestrian culture defined by status rather than nationality. The Spanish horse, developed from desert stock, combined strength and presence to become a highly desirable animal, favoured by kings and military leaders. With the flowering of the Renaissance, the development of riding as an art drew on the qualities of the Spanish horse as a means of noble display. An equestrian culture was born wherein superiority was proclaimed not through skill in battle but in the refined and rarefied atmosphere of the riding house.

From the early 1500s skill in the art of manége, the forerunner of today's dressage, was a defining feature of a nobleman in Europe. When Italian riding master Federigo Grisone founded his riding school at Naples in 1532, he began a fashion that swept across the Continent and before long training in the riding house became an essential part of the young nobleman's upbringing. There he would learn to enhance the natural movements of a stallion when courting a mare or challenging a rival into a stylized display of athleticism and virtuosity. The capriole, which means 'the leap of a goat' is probably the most famous and difficult of what were known as 'demi-airs' and 'high airs', depending upon their relationship to the ground. The horse leaps into the air above the height of a man and kicks backwards to the full reach of its hind legs. The execution of this movement has changed little since it was first performed, but the courbette of today, in which the horse bounds forward on its hind legs with its body almost vertical, was originally an exaggerated prancing movement known as a curvet, but has the same root meaning from the Italian word for a crow. The curvet was also the forerunner of the pesade, which effectually halted the prancing movement at its highest point so that the horse held a moment of suspension balanced on its hind legs.

As an illustration of the power of the horse under the control of the rider, Leonardo da Vinci, Peter Paul Rubens and Anthony van Dyck were among the many significant artists who chose this movement to enhance equestrian portraits of kings and great noblemen. While to the non-rider this pose may suggest that the horse is rearing in resistance to the rider or, to the modern follower of classical dressage, that it is performing a levade, neither would be correct. The curvet and pesade illustrate the perfect balance and control of the horse's forward dynamic, unlike the dangerous instability of a rear, but are also a preparation for forward movement. The levade is a stationary movement balanced on the hind legs and is recorded by the Spanish Riding School as being developed during the nineteenth century,

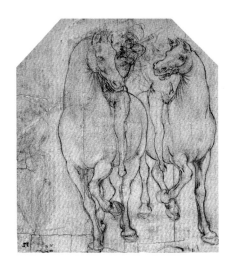

Preparatory study of two horses and their riders by Leonardo da Vinci.

so cannot be accurately applied to any work of art created earlier. This provides an insight into the way the perception of the audience can lead to several readings of the same material, and even during the long century of its high popularity, the semiotic value of the art evolved.

Initially it offered a metaphor for the control by man of his own animal passions and also supported the biblical imperative for man to have dominion over nature. Thus obtaining submission at any cost was acceptable, even if this meant tying 'a hedgehog' or some other 'criynge and biting beast' between the horse's back legs to force him forward, beating his head to make him stand still or ordering several servants to hold his head under water if he refused to cross a stream. Although the rider should use a 'milde and chearefull voice' and 'cherish his horse' for doing well, such common-sense advice was usually balanced by brutality. While 'the more you force him, the lesse he wyll yeald' makes a sound point, if all else failed, tying a cord around the horse's testicles for the rider to pull in the event of resistance was sure to do the trick.[2]

Terrible injury must have resulted for all concerned, and many of these forceful methods seem too ridiculous to ever have been put into practice. In contrast, with the rise of French riding-master Antoine de Pluvinel in the early seventeenth century, treating the horse with greater consideration became evidence of humanity. Pluvinel trained the sixteen-year-old King Louis XIII and believed a rider should know not only how to handle a horse but also how to be a truly noble man, secure enough in his supe-riority over a lesser creature to behave with compassion. Pluvinel's advice was never to ask 'more from a horse than half of what he is capable', and to remember that 'It is important that the pupil be a man and not a beast in man's clothing.'[3]

Royal interest in the art undoubtedly helped its success. In England, the sadly short-lived Henry, Prince of Wales, who died at just eighteen years old in 1612, was so devoted to horseman-ship that several books on the art were dedicated to him, includ-ing Gervase Markham's *Cavelrice: or the English Horseman* (1607) and Nicholas Morgan's *The Perfection of Horsemanship* (1609). Prince Henry erected the first riding house in England, and horses exceeded all other types of gifts given to him by foreign royalty and dignitaries. The giving and receiving of horses as gifts reflected their prestige among the nobility, while the ele-gant handling necessary to display their qualities became the measure of a man's own grace. Soldier-poet Sir Philip Sidney was taught that 'no earthly thing bred such wonder to a Prince as to be a good horseman',[4] and in his manual for gentlemanly education, Sir Thomas Elyot advises that

the most honorable exercise . . . that besemeth the astate of euery noble persone, is to ryde suerly and clene on a

great horse and a roughe, whiche . . . importeth a maiestie and drede to inferiour persones, beholding him aboue the common course of other men, dauntyng a fierce and cruell beaste.[5]

Elyot highlights very neatly several of the prime motivations in the elitist culture that developed around horsemanship. The importance of self-presentation had become a key feature in living in the context of the European court. The ability to acquit oneself as well in the riding house as on the battlefield showed cultured refinement alongside courage and strength.

Between the end of the English Civil War in 1646 and the restoration of the monarchy in 1660 many Royalist noblemen were in exile, uncertain if they would ever see home again. Among them was William Cavendish, Marquis (later Duke) of Newcastle, master of self-promotion and superb horseman. Newcastle set up a riding house in the former studio of the artist Rubens in Antwerp and opened his doors to the noblemen of

Europe. So great was his skill when he rode, sometimes with only a scarf around his horse's neck, that on one occasion some Spanish gentlemen among the crowd 'cross'd themselves and cried, Miraculo'.[6] Horsemanship served as a common language, so the nature of a man could be recognized by his handling of a horse, even if he was in every other respect a foreigner.

Newcastle also wrote two manuals that were, and remain, the only seminal texts on horsemanship ever produced by an English author. There were at least 20 different manuals dealing specifically with riding as an art in circulation around Europe from 1550 onwards. The forerunner, Grisone's *Gli Ordini di Cavalcare*, was translated into French, German, Spanish and English and ran to at least 22 different editions by 1609. The greater proportion of horsemanship manuals were derivative works, essentially tributes to a great master or his followers, with little claim to originality. Newcastle, however, was emphatic that 'my Book is stolen out of no Book, nor any mans Practice but my own', adding with characteristic self-confidence that 'it is the Best that hath been Writ yet'.[7] While he was part of a tradition of riding, his claims to originality were not unfounded and he is responsible for, among other innovations, the movement today called the 'shoulder-in', a standard exercise in the development of flexibility in the horse. His first manual, *La Méthode Nouvelle et Invention extraordinaire de dresser les Chevaux* (1658), with 43 plates of exceptional quality by Abraham van Diepenbeeck, was published in French for the Continental rider, and set out to gracefully supersede the methods of Antoine de Pluvinel, whose manual had appeared posthumously in 1623. Newcastle's second manual, *A New Method, and Extraordinary Invention, to Dress Horses* (1667), was a different, though closely related, text which openly undermined the continued reliance by English riders upon the methods of Grisone, by then over a hundred years old.

Published at his own expense, Newcastle's manuals largely followed an accepted progressive pattern of training from the first ride to advanced moves, and were both offered for sale and distributed as a gift among his friends, peers and the elite of the Continental courts. They were translated into French, German and Spanish, as well as freely adapted by several authors, but the best known edition is the 1743 translation of the original French text into English in a fine edition by John Brindley with all the plates, which made Newcastle's most significant work widely available for the first time.

By the end of the seventeenth century horsemanship as a cult of the nobility had lost popularity in favour of racing, but it had not died out altogether. In the eighteenth century Françoise de la Guérinière wrote perhaps the most famous horsemanship manual, *Ecole de Cavalerie*, finely illustrated by Johannes Elias Ridinger, continuing the tradition of fine texts by skilled horsemen. The Spanish Riding School of Vienna, established in the late sixteenth century, remains world famous today, travelling widely to give displays of what is now known as 'classical riding'.

The basic skills of the classical tradition form the foundation of many contemporary riding styles and the modern sport of dressage is directly descended from the riding house. The desire to experience the physical power of the horse within the boundaries of human control and safety is best typified in this tradition where every movement is based upon the natural expression of the horse, yet has been shaped and stylized for human comfort. The nobleman on horseback offered a cultural mirror to both human and animal nature which reflected the rider in relation not only to his horse but also to his place in the society of the elite and the accepted hierarchy of living things.

The Spanish conquistadors took the essentials of this style of riding to America with them in the sixteenth century, introduc-

The noseband in this 18th-century image by Elias J. Ridinger was designed by William Cavendish, Duke of Newcastle, a century earlier. The reins were not meant to pull the horse's head down, like draw reins, but to be used subtly by the rider to encourage lateral flexion.

ing horses into the landscape for the first time since their pre-historic ancestors died out. Thus the equestrian culture of the Native American peoples of the Great Plains grew not from a long tradition or a fashion, but from an opportunity that changed the ways of a people who had lived on foot for genera-tions. In 1541 the conquistadors reached present-day Kansas, where the tribes of the Plains spent their lives following the buf-falo, carrying everything they possessed. By 1673, when French Jesuit priest and explorer Jacques Marquette travelled the lands west of the Mississippi, the Illini tribe were there to meet him on horseback. Once the nomadic Plains tribes embraced the possibilities of this new fleet-footed and sensitive animal for hunting and travelling across the vast expanses of the grass-lands, a new cultural tradition was created that spread the horse widely across North America.

Iron Teeth, a Cheyenne woman, recalls that

> My grandmother told me that when she was young . . .
> [t]he people themselves had to walk. In those times they
> did not travel far nor often. But when they got horses,
> they could move more easily from place to place. Then
> they could kill more of the buffalo and other animals,
> and so they got more meat for food and gathered more
> skins for lodges and clothing.[8]

When horses were introduced to a tribe, gradually the carriage of supplies and goods was transferred to them from the dogs that previously had carried baggage. Once the young braves had shown their mettle in riding the new creatures, others would take up the potential for ease of travel and hunting. By the time the braves had become the elders, riding would be common-place, so that the next generation would not know a life without horses. Horses were, therefore, treated with great respect, and the intuitive relationship between the Plains tribes and the nat-ural world made the sensitivity and intelligence of the horse the heart of a particular relationship.

A Buffalo Hunt, by George Catlin (1796–1872).

Unlike the subjection seen as the starting point by Grisone, or even the cool superiority of human over animal intelligence advanced by Pluvinel, the pride of the Native American horseman was explained as a shared understanding by Chief Plenty Coup of the Absaroka, or Crow Tribe:

> My horse fights with me and fasts with me, because if he is to carry me in battle he must know my heart and I must know his or we will never become brothers. I have been told that the white man, who is almost a god, and yet a great fool, does not believe that the horse has a spirit. This cannot be true. I have many times seen my horse's soul in his eyes.[9]

From the early nineteenth century, a new horseback culture began to develop alongside, and ultimately at the expense of, the native people: that of the cowboy. While celebrated widely in literature from the work of Owen Wister, through Zane Grey and Louis L'Amour, to contemporary writers who paint a less romantic image, such as Larry McMurtry and Cormac McCarthy, the cattle-hand on horseback has been promoted most widely through the rise of film and television. Audie Murphy, John Wayne and Clint Eastwood are just a few of the names who made the horse, the six-gun and the Stetson hat lasting images across the world.

The relaxed riding style of the cowboy, so familiar from countless films, was rooted in the formal past. The arched neck and prancing hooves favoured by the horsemen of the conquistadors may have looked splendid on the battlefield but this style was impractical on the Great Plains of America. Gradually, while the deep-seated saddle and long leg position remained, the horse's head carriage moved from arched and collected to long and low. The high movements relaxed into a ground-covering walk, a gentle jog that would not alarm stock and an easy lope that could be maintained over long distances. The high pommel of the classical saddle, designed to keep a soldier in his

An elderly tribesman and his old horse remember past times. The horse is wearing a war-bridle – a single fine rope or rawhide strap around its lower jaw.

seat in the crush of battle, changed to the horn a cowboy would wrap rope around when lassoing cattle. The Spanish horse in the Americas served a role it had never known in Europe as the vast expanses of the landscape demanded different skills from both horse and rider.

The era of the great cattle drive, an iconic image of cowboy life, lasted less than 30 years but over 10 million head of cattle were moved from Texas northwards between the end of the Civil War in 1865 and the closing decade of the century. When new markets were needed for selling cattle in 1866, Charles Goodnight, a former scout for the Texas Rangers, and his partner, Oliver Loving, along with eighteen cowboys, drove 2,000 cattle all the way to New Mexico. They sold half the herd there, then drove the rest on to Colorado, fattened them up on the good grass there and sold them in Denver. Indian agencies and military outposts needed regular meat supplies and the possibility for government contracts opened up at once. But it was a dangerous journey. Holding the herd steady while they crossed rivers put the men at risk, while stampedes and attacks from hostile Comanche whose hunting grounds they crossed were a

Singin' em to Sleep, 1926. Western artist Frank Tenney Johnson grew up in a farming community in Iowa. His belief that horses are as individual as people makes them central to his images of the cowboy life.

constant threat. Loving had a reckless streak and when he insisted on riding on ahead to bid for a new contract, he was caught in the open by Comanche warriors and shot. He held his attackers off for several days before he escaped but was badly wounded. He refused the amputation that might have saved him until he could speak to Goodnight, by which time gangrene had set in and it was too late. But before he died Goodnight assured him that his body would be taken home to Texas. Their story is revisited in Larry McMurtry's *Lonesome Dove* and has all the drama, enterprise and tragedy that have become so associated with Western novels and films that a basis in fact seems hard to believe. While Goodnight returned to Texas with Loving's body, the Goodnight-Loving Trail they had established was seething with cattle and cowboys, all heading north.

Trail-drive cowboys were usually poor, and included a potent mix of black, white and Hispanic men, with ex-soldiers and former slaves among them. But they had to work together regardless of differences for up to four months at a time and in gruelling situations. Most men had three or four horses, often in poor condition, like the men themselves, and just as used to doing what was asked and taking what was offered in order to survive. The dust was so thick that the 'drag' riders bringing up the rear of the drive would hawk up black sludge at the end of the day but never get their lungs free of it. A startled cow could cause chaos, especially at night, and being dragged to death during a stampede was the most feared and the most common death for a cowboy.

Yet there are lyrical accounts of life on the cattle trail. Teddy Blue, English born but American bred, recalled that 'After you crossed the Red River and got out on the open plains, it was sure a pretty sight to see them strung out for almost a mile, the sun shining on their horns', while Goodnight considered that:

All in all, my years on the trail were the happiest I ever lived. There were many hardships and dangers, of course, that called on all a man had of endurance and bravery; but when all went well no other life was so pleasant. Most of the time we were solitary adventurers in a great land as fresh and new as a spring morning and we were free and full of the zest of darers.[10]

Further south, the pampas grasslands of South America also lent themselves to a horseback culture, a hard life and ultimately a similarly significant impact on the men who grew from it. When wild horses and cattle roamed freely across the fertile landscape of the grasslands between the Spanish Río de la Plata and Portuguese settlements on the Brazilian plain, descendants of the conquistadors and the native peoples became known as 'gauchos'. The term derived from the word for 'vagabond', and the gaucho was often considered barbaric and uncouth. Living on the margins of society but shaped by extreme loneliness, harsh weather and adversity, he had little time or opportunity for refinement.

A gaucho's worldly goods often consisted only of a horse and a long curved knife that served as both tool and weapon. Gaucho culture grew up around the cattle trade, initially for leather rather than meat. Known as the 'wanderers of the Pampas', gauchos were famed for their hunting techniques with three hard leather balls tied to a rope and thrown to bring down

A cattle drive in a snow storm by Frederick Remington (1861–1909).

animals, which were then hamstrung. This method of hunting required particular horsemanship skills and the gauchos, like so many other essentially nomadic equestrian cultures, depended on those skills for survival in a vast open landscape.

By the early eighteenth century the gauchos were considered too great a law unto themselves. From rough cattle herders, they came to be seen almost as outlaws but their standing changed with the late nineteenth-century wars against Spanish domination. A resistance movement grew up that was led by caudillismos, charismatic leaders who drew together fierce bands of fighting men on horseback. The gaucho's toughness and skill became a vital resource and the marginalized figure, always treated with great wariness by polite society, became a folk hero.

The caudillismo inspired intense loyalty among a following based less on his cause than on his personality. These men were not linked to any formal army or institution, yet by virtue of the band that gathered around them they were titled 'General' and came to wield huge political and personal power. Their nativist spirit caught the imagination of journalists and writers and they became cultural heroes, appearing in newspaper articles and pulp novels as icons standing up for the loyalties and values of the past. Like Genghis Khan, the caudillismos succeeded in drawing together people across social boundaries but, while they called themselves revolutionaries, their motivations could range from a personal vendetta to political dissatisfaction, and the men who followed them often did so purely from loyalty to the charismatic leadership they offered.

A caudillismo soldier would have several horses, possibly a firearm, which might be a virtual antique, but always a lance and long knife, keeping links with the Spanish cattle traditions. Wealth was measured in animals and in the hard machismo culture mares were undervalued, selling perhaps for only one peso, when even a colt foal could fetch five pesos. Riding a mare was considered preferable only to walking, but then anything was preferable to walking and a man would mount his horse simply to avoid walking a few steps. The isolated borderland estancias were small private forts with massive bolted doors, walled compounds and stone corrals that could hold several hundred or even thousand head of animals.

The horses were essential because of the vast distances between properties, the great expanses of cattle ranges and the need always for a quick means of escape from a hostile reception. Horses also provided the central focus for community gatherings, where displays of roping, racing, branding and castrating were part of the entertainment. The skill involved was a way of maintaining status and position in a way that was a far cry from the genteel art of the European gentleman on horseback. But handling cattle had made horseback skills natural and essential to generations of gauchos and when these men joined the armed bands of the caudillismos, the lance and knife they used in their hard daily work in a lawless environment turned easily into weapons of war. The lives of the herdsmen were mobile and harsh, honed by the competition for work and the struggle to survive in a dry land. Unmoved by hardship or suffering, these men used long steel barbs on their spurs which made their horses bleed, and regarded their own scars, often from bloody confrontations for the smallest perceived insults, as prestigious. Their resilience and spare needs alongside their riding and fighting skills made them very desirable as soldiers. These were common features among the Latin American borderlanders, Argentine gauchos, Venezuelan llaneros and Mexican vaqueros, who shared a similarly hard existence.

Today in Argentina, which covers 4 million square kilometres (1.5 million square miles), over 150,000 gauchos manage

55,000,000 cattle, 25,000,000 sheep and over 2,000,000 horses. From their early marginal status, the gauchos have risen to occupy a position linked with a fierce cultural pride in their traditions. Commodore Juan José Güiraldes, President of the Argentine Gaucho Confederation, describes the gaucho in terms that illustrate the complex, almost chivalric, code which has come to define a way of life:

> The Gaucho has solid principles. He believes in the given word and is true to his friends and to himself. He is austere. He has a definite concept of superior and subordinate. He is extremely patriotic and removed from politics. He imparts aesthetic beauty to handicrafts, silver work, weavings, braided leather and carvings in horn and bone. He uses language in a style of his own, in his stories and tales around the fire. He is a poet and musician, author, interpreter and dancer. He respects women and is sober in his love. Above all he has, and practices, a code of honour and conduct impossible without freedom. He has something that belongs to special people – a style of movement that implies aesthetics, manners and respect. He is proud of who he is.

Daniéle Da Meda's photographic series, *Gauchos* shows traditional tools in modern hands. The images were taken in the southernmost state of Brazil, Rio Grande do Sul and explore the region's socio-cultural heritage in the modern world.

These high ideals survive due to a strong sense of cultural identity, so that the horsemen of Argentina today 'search for and find ourselves in the Gaucho'.[11]

A significant similarity between all these equestrian cultures is that they primarily feature male riders and owners, even though women have a long tradition of association with horses, going back to the days when Epona's images were carved into the earth. However, the idea that riding is a male preserve was the dominant perspective in the Western world right until the turn of the nineteenth century, when riding astride became acceptable for women. In Eastern cultures, however, the traditional wide-legged trousers worn by women meant that riding astride had never been necessarily encumbered by skirts. Men did not always wear trousers, of course, and the simple tunic favoured by the Greeks and the Romans right through to the European middle ages must have made riding a breezy experience. Xenophon addresses the potential embarrassment of this when he writes that a rider should use his right hand to help himself leap elegantly onto his horse, so that 'he will not look ungraceful, even from behind'.[12]

Women then, were not the only ones who had clothing or the lack of it to hamper their progress on horseback. However, while the Sarmatian women of the lower Volga steppes, reputed to have been the inspiration for the Amazons in Greek mythology, were known to hunt and fight in battle alongside their husbands, the perception of war and exploration as male endeavours seems to have tipped the balance where riding on horseback is concerned also. Women on horseback are frequently noted for exceptional qualities, such as Queen Isabella of Spain in the fifteenth century, who demanded a spirited horse before she was ten and went on to excel in the riding and breeding of fine Spanish horses, and, a hundred years later, Queen

Elizabeth I of England, noted for her love of the hunting field. Female travellers on horseback challenged contemporary perceptions by setting out on ambitious journeys with little protection or support. Celia Fiennes undertook two long county by county journeys through England and Scotland on horseback in 1697 and 1698, riding side-saddle with only two servants to accompany her, while, in the nineteenth century, the intrepid Isabella Bird explored Australia, the Rocky Mountains, large areas of Asia and finally Morocco, when she was 70 years old, riding a black stallion given to her by the Sultan. Reckless riding can denote an especially feisty female spirit, as in Zane Grey's western romance, *Wildfire*, of 1917. His heroine Lucy is 'like a wild horse – free, proud, untamed' and rides a horse that 'ain't safe even for a man'.[13] Women riders in the circus or Wild West shows were similarly notable because they demonstrated courage and athleticism and took alarming risks.

Young Indian women play polo riding astride in this Rajasthan-style image with gilding on paper, c. 1750.

This focus on the danger of riding for women arises frequently and in 1890, James Fillis, *Écuyer-en-chef* to the St Petersburg Cavalry Riding School, wrote that:

> The great want in a man's seat is firmess, which would be still more difficult for a woman to acquire if she rode in a cross-saddle, because her thighs are rounder and weaker than those of a man. Discussion on this subject is, therefore, useless. Ladies who ride astride get such bad falls that they soon give up this practice.[14]

However, this view may reflect more upon deep male insecurities than on the problems of riding with rounded thighs. Indeed, the position of the lady's thighs was at the heart of the argument. Regardless of the skills and dangers of riding side-saddle, arguably greater than those of riding astride, so long

In a secure riding position, a lady can feel confident enough to flirt a little with the lower classes. From *Punch*, 1892.

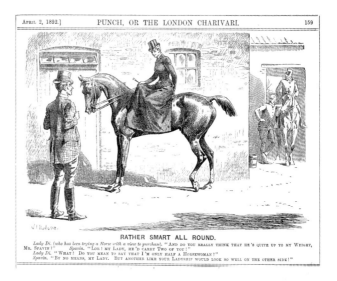

APRIL 2, 1892.]   PUNCH, OR THE LONDON CHARIVARI.   159

RATHER SMART ALL ROUND.

Lady Di. (who has been trying a Horse with a view to purchase). "AND DO YOU REALLY THINK THAT HE'S QUITE UP TO MY WEIGHT, MR. SPAVIN?"   *Spavin.* "LOR! MY LADY, HE'D CARRY TWO OF YOU!"
Lady Di. "WHAT! DO YOU MEAN TO SAY THAT I'M ONLY HALF A HORSEWOMAN!"
Spavin. "BY NO MEANS, MY LADY. BUT ANOTHER LIKE YOUR LADYSHIP WOULD LOOK SO WELL ON THE OTHER SIDE!"

as women kept their knees together, men seemed to feel more at ease.

The close links between women and horses may only be seen as developing into a widespread cultural tradition in the West from the nineteenth century, when women became as active as men in the pursuit of the British fox, rather than being a notable minority. In 1878 the visit to England by the Empress of Austria, a passionate lover of the hunting field and a skilful horse-woman, made hunting fashionable among women, though her habit of carrying a fan as well as a whip on horseback lasted only a short time among her followers. In 1952, prior to her coronation, Queen Elizabeth II became the first British queen to review her troops on horseback since her earlier namesake in the sixteenth century, Queen Victoria's intention to do so having been thwarted by the Duke of Wellington.

With the post-war rise of riding as a sport and recreation which was no longer limited to enjoyment by the elite, women began to ride astride routinely and this became, over time, unremarkable. Women excelling at riding, however, stirred complex responses until relatively recently. During the arguments over women's desire to ride racehorses in mid-twentieth-century America, one jockey revealed the true nature of the protests, 'If you let one women ride in one race, we're all dead.' This fear had already been justified back in 1804 when Alicia Meynell beat two male jockeys in challenge races and one trainer was soon admitting that, 'Our horses seem to respond better to girls than men.'[15]

In 1981 a BBC documentary film called *The Englishwoman and the Horse* was subtitled 'The great British love affair in all its batty glory'. In just over a century, the relationship between women and horses had progressed from being perceived as an exception to a male tradition with a controversy over decorum

In spite of dull Care
The Breeches I'll wear

I'll have them by the Mace I Dr. Clare

Of the Breeches I Deny you a share

THE HIGH METTLED JOCKIES
Or the best at three Heats for the Breeches

and safety, to an accepted passion of peculiar 'Englishness'. However, it seems likely that women through the centuries for whom horses were part of everyday life would have taken issue with this view in every detail.

Today, women who ride cross a wide age, culture and class range, as well as the full gamut of equestrian activities, both work and leisure. They often promote a more sensitive understanding of the horse, which is linked less to gender than culture, but it seems that women have less cultural baggage to overcome than men in this respect. The contemporary role of women in equine veterinary medicine is widely acknowledged, but when Aleen Cust graduated from the New Veterinary College in Edinburgh in 1896, the Royal College of Veterinary Surgeons could only license 'persons' to practice and, as a women, she did not fit into that category. It took until 1922 for her to become the first woman member of the RCVS so, in the meantime, she worked beyond their remit, in a village in Ireland, where she scandalized the local Roman Catholic priest by gelding horses. Even the most deeply ingrained prejudices can relax given time and equestrian sport has provided a high profile arena for women to demonstrate their affinity with horses. Women took part in the equestrian section of the Olympic Games for the first time at the Helsinki Games in 1952 and today the horseback events are the only ones in which men and women compete on equal terms.

Since 1986 the President of the Fédération Equestre Internationale, one of the Association of Summer Olympic International Federations, has been a female member of a royal family, with HRH The Princess Royal of Great Britain being followed in 1994 by HRH The Infanta Doña Pilar de Borbón of Spain and in 2006 by HRH Princess Haya Bint Al Hussein of Jordan. Both Princess Anne and Princess Haya Bint Al Hussein are Olympic riders. Women joined the regular uniformed members of the Metropolitan Police Mounted Branch in 1971 and the Royal Canadian Mounted Police in 1974, only two examples from the many available of women on horses in the workplace. At the most practical level women can ride as readily as men, but the well-established social barriers crossed by modern links between women and horses have created an equestrian culture that is unlike any of its forerunners.

While changing times have meant that the reliance on the horse that was at the heart of early equestrian cultures is no longer relevant, it is notable how many of them have continued in some form into the present. The link to old traditions and cultural roots that the horse symbolizes still has value. Today, as modern communications offer the like-minded rapid contact across continents, the cultural impact of the horse continues to develop in new ways and across far greater distances than ever before.

5 Into the Valley of Death

Maybe I'll be court-martialled
But I'm damned if I'm inclined
To go back to Australia
And leave my horse behind.
Australian Light Horse soldier, First World War

One of the strongest images of a man on horseback is that of the mounted soldier. There is an irony in this as, while the stallion is often presented as an image of aggressive male ferocity, the horse in general is not an especially fierce or even brave creature. As a flight animal, its best defence is the ability to outrun most of its attackers. However, if it cannot escape, it will defend itself and even the most gentle mare will turn on any threat to her foal. When stallions fight over herd leadership or mares in the breeding season, conflict of alarming ferocity takes place and a horse is undoubtedly a creature capable of inflicting enormous damage with teeth and hooves. Within domestic herds or small groups of horses order is often maintained by biting or kicking, and horses have their natural friends and occasional enemies that they remember from meeting to meeting over considerable periods of time. Close proximity between horses that simply do not get on can involve squealing and lashing out, usually with the hind legs, a defensive action, rather than the front, which is more aggressive. Horses seen on their hind legs 'boxing' in a field are rarely engaged in a serious battle and this mock fighting is usually linked to keeping the place in the herd. A confident leader is rarely challenged and it is horses in the group who are uncertain of their security who challenge and may bully or intimidate others. This, of course, may be applied

Two-year-old colt Gaffizar play-fighting, and also demonstrating the natural movements upon which classical dressage is based.

to human relationships also and offers a model for understanding our impulse to war that suggests similarities between humans and horses.

However, horses do not seek to acquire land for the sake of power, or to dominate other horses, unless they feel that their place in the herd is threatened. Certainly, they do not enter combat as a group against other animals, or humans. For them to go into a battle requires a reversal of their natural instinct to flee from loud noise or unfamiliar, alarming activity with all the speed nature has provided. Yet the instinct to run with and stay close to the herd and to follow strong leadership may propel them forward also, regardless of their bewilderment or fear.

A horse galloping headlong into battle because there are spurs in his side and all the other horses around him are galloping at least carries his rider into the fray. For him to perform there beyond his instincts to leave at the earliest opportunity would require something more. Arabic treatises stress that the horse's confidence in his rider is the key to success in battle, an observation echoed in Native American tradition. Among the British military, the ability to inspire a horse to charge across the hunting field, leaping anything in its path, was long held to be the essential groundwork for a successful cavalry career. A horse that trusts his rider will face situations he would never dare to confront alone and, once on the field, that trust would have to remain firm enough to overcome all distractions.

While stallions may seem suitable as warhorses, since they are apt to respond to aggression with aggression, the calmer

Detail from the Bayeux Tapestry in Caen, Normandy, a contemporary depiction of the Battle of Hastings of 1066.

focus of a mare or gelding offers less fire but more manageabil-
ity. Aside from gender preference, choosing a bold or very
steady horse over one that is naturally sensitive or nervous
would be crucial for riding into any sort of conflict. This point
was recognized by Xenophon around 350 BC and remains the
reason that horses chosen today for work in, for example,
mounted police forces undergo so many tests of character
before securing a place. However, the warhorse and the horse
that is requisitioned for war may well differ in every way and
horses trained to pull carts, plough fields or carry children have
nonetheless died on battlefields in their thousands.

The earliest known images of domesticated horses, from the
Middle East around 2000 BC, show them drawing war-chariots,
even though archaeological evidence from the southern
Ukraine suggests that they had been ridden for at least 2,000
years longer than that. The work of daily life was of less use as
propaganda than images of battle, though this agenda can
make the accuracy of information doubtful. Certainly, images
of horses in battle present the might and glory of war leaders,
though the suffering and loss of the common soldier and his
horse may be recorded also, as in the Bayeux Tapestry.

While the 'airs-above-the-ground' still performed today in
classical dressage, such as the capriole and the courbette, are
theoretically derived from battle moves based on the natural
behaviour of stallions, their practicality in the crush of battle
was a matter of debate from the Renaissance onwards. In India
very similar moves had been devised from as early as the thirteenth century, some designed to raise a warrior high enough
to attack an enemy mounted upon an elephant, and certainly
all of these techniques would have looked splendid and
instilled alarm into anyone who found a horse in the air above
his head. However, aside from the time and cost involved in

training suitably athletic horses, whether they or their riders could pause in the face of a screaming assault to perform feats that invited a sword in the horse's exposed belly is questionable.

The glorious charge where man and horse unite courageously against a common enemy was an inspiration to a great many poets and artists, but George Bernard Shaw's Captain Bluntschli shatters the romantic illusions of Raina that the leader of the cavalry charge is 'the bravest of the brave' with the pragmatism of the experienced soldier:

> Hm! You should see the poor devil pulling at his horse . . . It's running away with him, of course: do you suppose the fellow wants to get there before the others and be killed? . . . You can tell the young ones by their wildness and slashing. The old ones come bunched up under the number one guard: they know that they're mere projectiles, and that it's no use trying to fight. The wounds are mostly broken knees, from the horses cannoning together.[1]

Scotland Forever! by Lady Butler (1846–1933). The Charge of the Scots Greys at the Battle of Waterloo, 1815.

Willing or no, the use of the horse in war changed the nature of human conflict. A warrior on horseback has many advantages over a foot soldier, not least the potential for 'shock combat', where the sheer weight and mass of galloping horseflesh is used to plough through the enemy. This tactic was used by armies as historically and culturally diverse as those of Alexander the Great and William the Conqueror. By way of contrast, horses can be used for rapid strikes or tactical retreats, a technique favoured by the Parthian cavalry archers who, while appearing to flee the battle, would turn and fire as the horse galloped away, hence the 'Parthian' or parting shot. A mounted man can also intimidate an enemy on foot, added height giving greater velocity to the swing of a weapon, and the fear of being run down by a horse in full gallop is enough to make the most courageous defender step aside.

Horses could serve as draft as well as riding or chariot-harness animals, immediately making every aspect of an army expedition more efficient. On horseback, the ambitious or the vengeful could

From a Chinese tomb dated between c. 200 BC and 200 AD, this is believed to be the earliest illustration of the Parthian shot.

A Japanese warrior, Sato Masakiyo, with his horse and page, attributed to Yoshitora Utagawa, *fl.* 1850–70.

藤
原
正
清

search further for land or retribution, while defensive and relief forces could arrive where they were needed far more quickly. A horse increased the mobility of an army by far more than the ratio between how far a man and a horse can travel in a day.

Many types of horses have been used in warfare, from the native breeds, such as the hardy steppe ponies of Eurasia, to the destriers bred to carry the weight of a medieval knight in full armour, although these were not the size of Shire horses, as is commonly believed, but more like a modern heavy hunter. War was one of the main means by which horses left their native countries and therefore is responsible for much of the introduction of new blood to existing stock. The Barb of North Africa is, along with the Arab, one of the oldest and most influential breeds and was introduced into Europe in large numbers during the upheavals of the eighth century. During the invasion of Spain the Berbers formed part of the Muslim armies and the horses they rode into battle went on to influence the development of the Spanish horse. The sort of horse used for war depended largely on fighting style but the heavier breeds favoured by European armies lacked speed. When the Muslim armies were defeated at the Battle of Poitiers in AD 743, the sturdy Frankish horses could not keep up with the lightly built Barbs as they made their escape. But enough of them remained behind to form the foundation stock for the Limousin, a purpose-bred charger, combining the strength and weight-carrying ability of the French stock with the greater endurance and speed of the Barb. In 1675 a Limousin mare named La Pie, the favourite mount of the Viscount de Turenne, Marshal General of the Armies of France, was said to have continued to lead her master's attack on the Austrian guns even after he was killed in the battle.

The value of light cavalry gained a particular reputation through the success of the Hussars raised by King Matthias Corvinus of Hungary in the fifteenth century. One man in every twenty in a village was required to take military service and they became famed for their rapid strikes, impressive mobility and

This Oglala war party, photographed by Edward Sheriff Curtis (1869–1955), illustrates the strong visual impression intended to raise the warriors' own morale and undermine that of their enemies.

rejection of the formalized set-piece battle. Corvinus knew just what he wanted in a warhorse and his rejection of 'horses that hop about with bent hocks in the Spanish fashion' must have been a blow to those who thought the Spanish horse the epitome of equine efficiency and elegance for all purposes. The Hussars seemed to prefer plain practicality and his ideal was 'horses that stride out and stand firm when required'.[2]

The nature of the horse was as important as its physical type and the Hindu merchants of twelfth- and thirteenth-century India had an equine caste system, based upon character. A Brahman could be relied up to be brave in battle, while a Khsatriya was known for endurance. A Vaishaya was likely to shy away from anything that frightened it, while the Shudra was a coward who would throw a rider and head for safety. Horses trained on the hunting field were recommended as bold and unflinching in war by Xenophon, and a nature that rose to challenging sports was long considered the best foundation for a warhorse.

Once a certain breed or type of horse became popular for cavalry use, demand rose and the strict management of suitable types was common during the centuries when the horse was an essential component of successful warfare. Between 1279 and 1282, Phillip III controlled the breeding and movement of warhorses carefully to ensure sufficient provision for France. Landowners had to keep at least four brood mares and breeding stock were not to change hands in the settlement of debts. He imported heavily but forced any merchant with a surplus of warhorses to forfeit them and forbade the movement of suitable stock out of or even through the country. Similar patterns emerge elsewhere, with the specific type of horses needed for war being influenced not only by the preferred battle tactics but also the climate. In India hot-blooded horses were preferred as chargers during the sixteenth-century era of the Mughal Empire, with fine coats, thin skins and hard hooves to enable them to cool down swiftly after exertion and cope with the challenging terrain. Cavalry horses were graded into seven types depending on their usefulness and expendability and all export of horses was forbidden. Over time and across the world the horse has not only been ridden into battle and judged on its performance there, but shaped in body and mind to become more suitable for the purposes of war, a fact for which humans should surely feel ashamed.

Most of the equestrian cultures discussed in the previous chapter have been involved in warfare in some way during their history. While Genghis Khan's cavalry existed to wage war and bring his vision into being, the American cowboy, ostensibly a cattle-herder, is equally famous for his range-wars, encounters with native tribesmen and gunfights. The Crusader knight, fighting his way across two centuries, is possibly one of the most familiar images, while the Muslim warrior who rode against him was an equally successful and formidable force. Relying on completely opposite tactics, when the heavily armoured knights of Richard the Lionheart, riding solid European stock, met the lightly protected warriors of Saladin on their nimble Arab horses in the late twelfth century, their conflict devastated both sides and led to an obsessive fascination between the two leaders. When Richard's horse was killed under him in battle, Saladin, rather than taking advantage of his vulnerability, sent him a fresh horse. History does not relate what sort of horse Saladin provided for his enemy but the choice would not have been made at random.

Among the Arab warriors of the medieval period, of whom the majority were cavalry, a tribal system operated whereby cavalry were supplied as and when required by the ruler caliphate. Under Islamic law, a warhorse was entitled to a share of the spoils of war with his master, with purebreds being entitled to a larger share than crossbreds, and stallions and mares receiving higher portions than geldings. This shows firstly that both entire and neutered stock were ridden into battle, but also that the breeding of high quality horses for the cavalry was actively encouraged by the rewards offered. Also, as a cavalryman received a higher portion of the bounty than an infantryman, aspiring to the cavalry had many incentives. These rules varied a little across the centuries and under different rulers but, broadly speaking, it was worthwhile to become a cavalryman with the best horse possible. There were downsides also: if a horse died while in army service, but not in battle, the rider had to finance his own replacement, and if they both died prior to the battle, his family received nothing and some of the benefits of his service were taken away from them. However, if the rider successfully killed an assailant in battle, he was entitled to that man's horses, personal armour and weapons. As discussed in

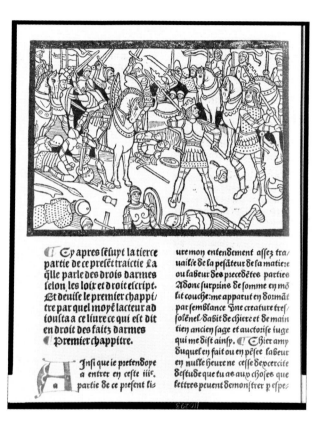

The king fights hand-to-hand among scattered body parts around the feet of his cavalry horses in this woodcut, c. 1489.

¶ Cy apres sensupt la tierce
partie de ce present traictie La
qlle parle des drois darmes
selon les loix et droit escript.
Et deuise le premier chappi/
tre par quel moyé lacteur ad
ioulta a ce liure ce qui est dit
en droit des faitz darmes
¶ premier chappitre.

A Insi que ie pretendoye
a entrer en ceste iiii.
partie de ce present li=

ure mon entendement assez tra/
uaillie de la pesanteur de sa matiere
ou labeur des precedétes parties
Adonc surprins de somme en mó
lit couche:me apparut en dormát
par semblance vne creature tres/
solénel Sabit de chiere et de main
tien ancien sage et auctorise iuge
qui me dist ainsp. ¶ Chier amp
duquel en fait ou en pésee labeur
en nulle heure ne cesse de percise
destude que tu as aup choses que
lettres peuent demonstrer p espe=

chapter Three, the purity of the Arab horse's bloodline was always highly valued in its native home and this is in direct contrast to the European warhorse, which was bred and cross-bred with a specific type in mind, namely a heavy weight-carrier that was up to the burden of a man in full armour.

The mounted knight, perhaps the most familiar image of the soldier on horseback, is linked with the romance of Arthurian legend, appears in many fairy tales and myths and remains

popular in modern fantasy fiction. But this warrior was very much a historical presence and the practicalities of mobilizing a man in full armour made preparation for war far from romantic. While the Arthurian knight gallops through fine art with only his horse, his sword and nobility of soul to support him, despite early vows of poverty, the elite forces of the Knights Templar did not believe in travelling light. Over one hundred of their complex rules relate to the supply of horses and the basic requirement was that senior officers should have four warhorses and a general riding horse. The Grand Master of the Temple also had a personal chaplain and a clerk with three horses, a sergeant with two more and a valet with yet another to carry his shield and lance. He was entitled to a farrier, an interpreter, a cook and a native-born man-at-arms on horses and a couple of young lads on foot for fetching and carrying duties. An optional extra for the

Master was a Turcoman horse to be kept in the caravan, that is, as part of the general pool of remounts, and led into battle by a squire to stand in reserve. This was a desert-bred horse, thought to relate to today's Akhal-Teke of Turkmenistan, which combined strength and endurance with the ability to survive on short rations of food and water. It was also one of the options for general riding, so seems to have been used as an emergency back-up horse in various circumstances, chosen for its naturally robust qualities.

The Crusades also saw the gradual improvement of armour for the horse as well as the rider. The warrior's horse would always have needed protection and the earliest form was reinforced leather padding for chariot horses of the Near East and Egypt around 2000 BC. Armour including head protection has been found in archaeological remains from Central Asia from the sixth century BC and protection of overlapping scales of leather or metal attached to a fabric base became common throughout the ancient world. Greek and Iranian reliefs showing horses wearing armoured panels date from around 200 BC up to AD 500. The Romans favoured foot soldiers over cavalry, but did have specialized units of armoured cavalry during the third and fourth centuries AD. However, it was not until the late eleventh century that horse armour became common in the armies of western Europe.

By the sixteenth century the armour of horse and rider had developed into a complex matching ensemble, called a garniture, with elaborate designs etched into the surface of the metal with acid. These were exceptionally beautiful in many cases, requiring great artistry of construction and decoration, with scrolls and foliage alongside the coats of arms and inscriptions engraved as part of the craft of the armourer. Specialists in the art had their own workshops and the prestige of armourer and

French dragoon
soldier, c. 1806.

client was involved in this accoutrement, though some of the most elaborate designs were intended more for ceremonial and tournament purposes than the battlefield.

The use of heavy armour for horse or rider declined even as it reached the height of artistry, becoming outdated by new weapons and battle tactics, so that by the beginning of the seventeenth century it already had the romantic appeal of a bygone age it retains today. Whether it was of any great use or comfort to the warhorses that laboured beneath its weight is debatable and it seems unlikely that beautiful etching would have made much impression upon them. Despite the fashions of various times, the horse still ended up in the middle of human conflict, protected or not.

This inevitability has led to horses dying in war in astonishing numbers. In 1448 at the battle of Caravaggio 10,000 horses were killed. In the 1812 Peninsular War one division alone of Napoleon's soldiers in Russia lost 18,000 of 43,000

horses in two months and during the ignominious retreat from Moscow 30,000 horses died, primarily from cold and malnourishment. The Duke of Wellington constantly requested remounts and wearily described the Peninsular as 'the

The Charge of the Imperial Hussars, by Jacques François Swebach-Desfontaines, 1820.

grave of horses'.[3] His 14th Light Dragoons lost 1,564 horses of the 1,840 at their command, against losses of 654 men. In the Boer War 500,000 horses, mules and donkeys were killed, while the staggering figure of 1,400,000 are estimated to have died during the American Civil War. Out of the million horses serving the British Army in the First World War, 500,000 were killed. In the Second World War the German army lost an average of 865 horses per day over 2,000 days, with over 52,000 at the Battle of Stalingrad alone. In 1939 the Polish Cavalry lost 2,000 horses in 30 minutes during the Siege of Warsaw and the road into the city was lined with dead and dying horses.

A postcard from World War I.

PALS !

Against the vastness of these numbers lie individual accounts from soldiers which frequently show that the importance of the horse to its handlers went far beyond its usefulness in battle. A Royal Artillery Officer in the First World War recalls the impact of having a favourite horse declared unfit for duty:

> The vet insisted Sailor was old, too thin and unfit for service. He didn't know. He only judged for appearances. Sailor wasn't much to look at, but was worth six horses in the battery. If a gun team jibbed, we hitched up old Sailor and he pulled them through. If a vehicle got stuck in a ditch, or was too heavy to start, old Sailor moved it. He would work for twenty-four hours without winking. He was quiet as a lamb and as clever as a thoroughbred, but he looked like nothing on earth, so we lost him. The whole battery kissed him goodbye and the drivers and gunner who fed him nearly cried.[4]

Seeing horses die caused a complex distress and Lieutenant Tom Butt of the King's Own Yorkshire Light Infantry recalls that while it was 'shocking seeing one's friends of all ranks lying dead . . . the sight of dead horses sickened us even more. They were not free men like us.'[5] The sense that horses had no place in the conflict is echoed by William 'Nobby' Clarke of the Queen's Bays:

> We were often hungry, and so were the mules and horses and how those poor creatures suffered. I think they must have been more tired and out of condition than we were. Innocent victims of man-made madness. They broke your heart, especially when you passed the

injured ones left to die, in agony and screaming with pain and terror.[6]

Private Sydney Smith seems haunted by his memories of 1917:

I had the terrible experience to witness three horses and six men disappear completely under the mud. It was a sight that will live for ever in my memory; the cries of the trapped soldiers were indescribable as they struggled to free themselves. The last horse went to a muddy grave, keeping his nostrils above the slush to the last second. A spurt of mud told me it was all over.[7]

At the beginning of the end of the American Civil War the surrender terms drawn up at Appomattox Court House, Virginia, in 1865 insisted that every Confederate Army cavalryman be

Australian Light Horse soldier and his horse, in a preparatory drawing for a magazine cover, by Edward Penfield, 1915.

allowed to take his horse home with him. However, the end of a war for horses very often meant not returning home to an honourable discharge, but being left behind, with a bullet in the brain if they were lucky.

After the First World War horses serving with regiments in the desert campaigns were sold to live out their days in hard labour as cart and pack horses on the streets in countries where the people barely had enough to feed themselves, let alone a horse from a different native environment, ill-suited to the fierce heat. Some officers shot them rather than subject them to this future, while the Australian Light Horse, based in Palestine, simply refused to comply. Learning that their horses were to be sold either as working stock or slaughtered for hides depending on their age, they arranged a brigade race meeting. Then the next day, they took the horses out, gave them a nose-bag and, while the horses enjoyed a last feed, shot them all.

A young dispatch rider and his pony, c. 1914.

When author Richard Holmes decided to recreate the retreat of British Expeditionary Forces from Mons to Marne between August and September 1914, he bought a horse called Thatch for the journey, planning to sell him afterwards. But he kept him because,

> After all that endless grooming, feeding and tacking up I began to understand how the cavalryman felt about his steed. I now know why, among the all the acts of petty governmental meanness at the war's end, the decision to sell the Army horses locally – rather than spend money shipping them home – caused such resentment.[8]

By the 1930s the contribution of the horses that died in the conflict was valued so highly that a memorial was proposed in their honour. However, Mrs Dorothy Brooke, based in Cairo, where her husband was a Major-General, wrote a letter to the *Morning Post*, later *The Daily Telegraph*. She began by offering an alternative means of honouring these horses and went on to explain:

Casualties of World War I: dead horses lying by the road-side.

The changing technology of war: cavalry horses in 1914 wearing masks with breathing tubes during a mustard gas attack.

Out here, in Egypt, there are still many hundreds of old Army Horses sold of necessity at the cessation of the War. They are all over twenty years of age by now, and to say that the majority of them have fallen on hard times is to express it very mildly. This country, to begin with, is not suitable to our horses: the heat, dust, want of water, and the fact that European horses are bigger framed and require more food than the poorer class of owner is able to supply, all add very much to their sufferings. Those sold at the end of the War have sunk to a very low rate of value indeed: they are past 'good work' and the majority of them drag out wretched days of toil in the ownership of masters too poor to feed them – too inured to hardship themselves to appreciate, in the faintest degree, the sufferings of animals in their hands.[9]

Mrs Brooke went on to appeal for funds to buy these ailing former cavalry horses and her emotive yet reasoned argument

A soldier comforting his injured horse: a statue by Henri Désiré Gauquie (1858–1927) in honour of the 58th British 'London' Division at Chipilly in Picardy, northern France.

resulted in donations with the modern equivalent of almost £20,000. She bought 5,000 aged cavalry horses and set up the Old War Horse Memorial Hospital in Cairo in 1934, promising free veterinary care for all the working horses and donkeys of the city. Today, the legacy of her work continues through the world's largest charity for the welfare of working horses, mules and donkeys. The Brooke offers veterinary care and support in Asia, Africa, Central America and the Middle East to ensure healthy working animals for people who depend upon them in some of the poorest communities of the world.

A horse, with its instinctive fears of noise, crowding, blood and chaos, has no place on a battlefield, and the days when horses had to suffer this regularly and in their millions are in the past. The First World War was the last major conflict to use regular horse cavalry in such numbers. However, while today the image of the horse in war may be seen as primarily historical, for all the modern technology at the fingertips of today's warrior, horses are still called upon to serve occasionally. In the wild mountains of Afghanistan, horses were the only means by which the Afghan Northern Alliance could attack the strongholds of the Taliban and in 2001 US Special Forces found themselves part of a cavalry unit of 300 when motorized vehicles became impossible due to roadblocks. The role of the horse in war is, sadly, not yet over.

However, it is greatly diminished and that is as well for the horse. The complex ethical dilemmas involved in taking an animal into war are swept aside once the enemy is perceived as mutual. However, the response of soldiers to their horses and

Time this old war horse was taken home; his 'tucked up' belly and hindquarters suggest he is cold as well as under-nourished and weary. By Edwin Forbes, 1863.

the suffering they experience through human conflict highlights a deeper responsibility. As humans find far more efficient means to kill one another than the cavalry charge, the way in which we look back on the slaughter of earlier ages can change too. It is hard to see the charge as glorious when men wept to watch their horses die, but the ability to witness the tragedy is what matters most. Today, when we can press a button to kill people on the far side of the world, perhaps we are in greater danger than in the days when we were forced to watch men and horses die side by side.

6 From Breadwinner to Performer

They always seemed to think that a horse was something like a steam-engine, only smaller.
Anna Sewell, *Black Beauty*

In *Tess of the D'Urbervilles*, published in 1891, it is the death of a horse that precipitates the tragic life of Thomas Hardy's ill-fated heroine. Prince, 'only a degree less rickety' than the laden cart he pulls, is killed when he is speared by the shaft of an approaching mail-cart as he plods the road in half-light with Tess asleep at the reins. The contrast between the slow journey of Prince, a bewildered, emaciated animal 'lacking energy for superfluous movements of any sort' and the mail-cart 'with its two noiseless wheels, speeding along . . . like an arrow' highlights the changing world of the nineteenth century. Prince is a family member, hungry as they are hungry, his name reflecting the shiftless father's delusions of nobility. The mail-cart horse is not described at all, except as having been uninjured. While Prince bleeds to death in a ditch, the mail-cart horse disappears, machine-like, into the future. The burden of responsibility Tess feels illustrates the reliance of her family upon the horse and the narrative comment, 'The bread-winner had been taken from them: what would they do?' sums up the importance of that life-line for travel, work and farming.[1]

The tension between the country and the city that grew during the long century of the Industrial Revolution was frequently addressed through the changing role of the horse. While the impact of industrialization would be felt worldwide, Britain's

The cart-horse as a symbol of patience and compassion in a children's book of moral stories, 1879.

THE HUMANE CART-HORSE AND THE CHILD.

enthusiasm for this radical change led the way and influenced Europe, Asia and America. The potential for such rapid development in Britain was linked to the accessibility of coal supplies and cheap labour, the import of raw material from the colonies and the resulting deep pockets of the merchants who could provide investment capital. From being an essential feature of the working environment, the horse would, slowly but inevitably, become an anachronism, simply no longer strong or fast enough to keep up.

Although *The Pickwick Papers* was published over 50 years before *Tess of the D'Urbervilles*, Charles Dickens considered the same frictions. The richness of an August cornfield, so ripe with growth and fruitfulness that a 'mellow softness appears to hang over the whole earth', is placed in contrast to a coach, cutting through the peaceful landscape at a cracking pace without regard for the bountiful surroundings. The workers in the field stop to witness the novelty of it, their skin tawny from the sun and their children shrieking with delight. Among them,

> The reaper stops in his work, and stands with folded arms, looking at the vehicle as it whirls past; and the rough cart horses bestow a sleepy glance upon the smart coach team, which says, as plainly as a horse's glance can, 'It's all very fine to look at, but slow going, over a heavy field, is better than warm work like that, upon a dusty road, after all.' [2]

The rough carthorses, like Prince, are part of the world receding slowly into the past but the true countryman and his horses share a wisdom that those active in the new ways have no time to notice. This appreciation of the slow pace of country life and the quiet pride in a simple lot was not only a romanticizing of what was a very hard way of life. Along with, and perhaps in reaction to, the changes of the Industrial Revolution, at the end of the eighteenth century there was a great sense of pride in farming as a craft, with ploughing matches being a very popular way of maintaining high standards. Anticipation of decline perhaps heightened appreciation of ways too traditional to warrant much notice until they were under threat, but they started to be seen at variance with the growing need for efficiency and output.

For both Dickens and Hardy the working pace of horses illuminates the change and by 1946 John C. Milne's poem 'Nae
nowt for me' attempts to brake the rapid forward movement
that was outstripping them altogether. His young Scottish lad,
whose yearning is for horses over anything the modern world
can offer, aspires only to work the fields with his own matched
ploughing-team:

> Big fite horses wi' silken hair,
> And milk-fite mains ahinging doon
> Fae their smooth curved necks like a bridal goon.

His quiet ambition is full of anticipated pride and he boasts that
he will feed his horses only the best and make sure they sleep at
night 'On fresh clean beddin twa feet deep'.[3]

The nostalgia for heavy horses, labouring alongside their
contented owners to make hay while the sun shone, began even

Early photograph
of plough-horses
by seminal land-
scape photographer
Frank Meadow
Sutcliffe. The horses
are Cleveland Bays,
a breed with a long
working history, but
on the Rare Breeds
Critical List in 2008.

as their role changed. Yet change it did, as many of those who worked the fields desired the better wages and higher standard of living promised by burgeoning technology. Many societies for the promotion of the heavy horse were established in Britain during the second half of the nineteenth century, such as the Clydesdale Society and the Suffolk Horse Society in 1877 and the Shire Horse Society in 1878. While their work in the fields would continue at varying levels until the Second World War, they were given new roles in the growing cities too. In Hardy's novels it is the tragic lives of humans that cast a shadow over the summery fields and in those of Dickens human greed turns the crowded cities into slums. The life of the working horse in town or country was decided entirely by the prosperity and attitudes of its owner and the increased demands of the changing times.

In 1798 Robert Bloomfield drew attention to the sufferings of those horses that sped by in a blur too fast for country folk to see. The post-horse was not governed by the rhythms of the seasons but 'Hired at each call of business, lust or rage' that 'prompts the travelling on from stage to stage'. Rather than a valued working-companion, the post-horse was simply a vehicle, ridden into the ground and then left in a cold stable where 'every nerve a separate torture knows'. Barely rested, he was brought out again to be whipped down the road to his next stop until 'His piece-meal murderers wear his life away'.[4]

The post-horse was the early equivalent of the hire car and while the ability to ride among all but the lowest working classes would have been common, the standard of horsemanship would have been very varied and those for whom a high level of riding was part of their education would have rarely needed to rely on hired horses. In Leigh's *New Picture of London*, a gazetteer of the city printed in 1819, there are over

60 places listed where a post-horse could be hired, with the close proximity of several suggesting that competition for custom would have been high. Frequent changes of rider, with doubtful abilities and places to get to in a hurry, must have meant that the life of even a well cared-for post-horse had little to recommend it.

Yet the earliest developments of the Industrial Revolution depended on horse-power. When Jethro Tull, the pioneer of agricultural improvement, published *The New Horse Hoeing Husbandry* in 1731, he set out to make farming with horses more efficient, not to rule it out altogether. But with developing mechanization the pressure to rival emerging forms of new technology meant that hard times were approaching for all working horses. While nineteenth-century Britain was frequently described as 'hell for horses', the term appears in print as early as 1621. As people teemed from the country into the cities during the extended century of the Industrial Revolution, hell became a very crowded place indeed.

Hardy's neat mail-cart would have been the local representative of a vast enterprise to speed and standardize the delivery of mail in Britain. The coaches of the Royal Mail were pulled by teams of drivers and horses from the main public houses along the route. While the landlords would keep several teams of their own horses for work in and around the city, once the coach left and headed out to other areas, teams were subcontracted from inns on the stages along the road. Among the most successful owners was William Chaplain, who owned 1,800 horses by 1838, when his business was at its peak, and supplied teams for over half the Royal Mail coaches of London. He had stables on all the most important roads leaving the city, one of which could accommodate 150 horses. An efficient ostler could have the change from one team of horses

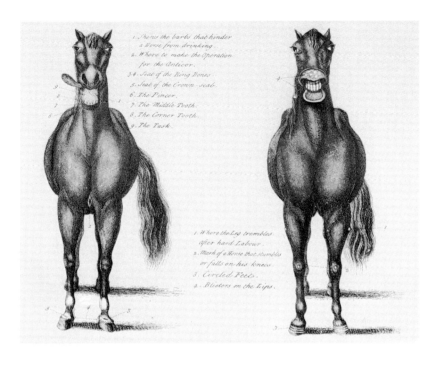

1. Shews the barbs that hinder a Horse from drinking.
2. Where to make the Operation for the Anticor.
3.4. Seat of the Ring Bones.
5. Seat of the Crown-scab.
6. The Pincer.
7. The Middle Tooth.
8. The Corner Tooth.
9. The Tusk.

1. Where the Leg trembles after hard Labour.
2. Mark of a Horse that stumbles or falls on his knees.
3. Circled Feet.
4. Blisters on the Legs.

to the next managed in less than three minutes, although a whole five was allowed, and his handlers would be at the ready upon hearing the post-horn announce the approach of the mail-coach. At the height of the mail-coach service there were over 150,000 horses in daily use. The preferred breeds were Cleveland Bays and other strong types of medium build and height with the attributes of both strength and speed. The stages were usually ten to fifteen miles apart and a grey horse was often used on the lead off-side to allow for visibility in poor light and at night.

In the cities the horse-drawn omnibuses kept people mobile in huge numbers, and in 1893 W. J. Gordon wrote:

A veterinary chart from the 1743 English translation of Gaspard de Saunier's *The Perfect Knowledge of Horses,* first published in French in 1734.

Study of a lying horse with raised head by Nicolas Sicard, 1881.

They have, in round numbers, ten thousand horses, working a thousand omnibuses, travelling twenty million miles in a year, and carrying one hundred and ten million passengers . . . every omnibus travels . . . sixty miles a day, and every horse travels twelve miles a day. And as an omnibus earns a little over eightpence-half-penny a mile, and the average fare paid by each passenger is a little under three-halfpence, it follows that each omnibus picks up six passengers every mile.[5]

The vast numbers of horses involved with the growing need for mobility raised concerns from the early eighteenth century and the treatment of horses came under scrutiny with the advance of veterinary knowledge. Up until then, the care of a horse's health was generally left to the farrier or the owner, who could find plenty of remedies recommended in horsemanship manuals and guides to animal husbandry. Most are of debatable value and some rather alarming. Among the recipes 'taken from the Best and most Modern Writers' by Sir William Hope

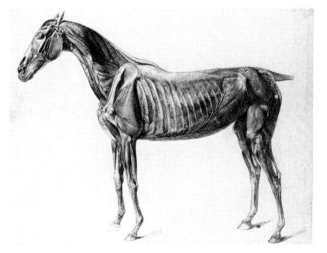

Published in 1766, *Anatomy of the Horse* by George Stubbs remains a seminal work, one based on the artist's own detailed dissections.

in 1696 is a remedy for a horse with stiff and tired legs, which begins, 'Boil six little puppies in Lees of Claret till their flesh be separated from their Bones', while the reliance on 'quick-silver', 'Hoggs Grease' and 'the Dung of a Goat, newly made' suggests witchcraft rather than medicine.[6] Blood-letting and purges were consistently popular, alongside, unsurprisingly, treatments for horses suffering from debilitating weakness.

By the beginning of the eighteenth century writers such as William Gibson were attacking these primitive treatments and the suffering they caused and it is perhaps this that represents the most significant step forward. At a time when the remedy for a horse that insisted on putting its tongue over the bit was to sear it off with a hot iron, suffering was rarely considered. Wilful cruelty was censured for the poor light it cast on the owner rather than the suffering it caused the animal, so severe practices undertaken with good intent would be quite acceptable. As the great range of horse-drawn vehicles struggled to keep up with the demands of burgeoning human development, fire-engines, hackney cabs, delivery wagons and carriages were all horse-drawn, some by up to four horses, tearing, plodding or high-stepping through the streets. The huge numbers burdened the already groaning resources as the need for stabling, feeding and the resulting waste built up. Collisions and fatal accidents involving horses were frequent, as were horses collapsing under impossible loads, quite literally worked to death on the streets in full view of the public. Suspicions were raised too that their remains were fed not only to cats and dogs, but were also surreptitiously introduced into human food as sausages. Horrifying tales of knackers' yards emerged, where horses were kept without food and drink alongside the carcasses of their fellows while they awaited slaughter. As the struggles of the working classes and their animals moved in from the picturesque distance, alongside the growing demands of technology and economics were stirrings of responsibility and conscience towards the less fortunate, both human and animal.

The Kill, a fox-hunting scene by Thomas Rowlandson, 1787.

This splendid hunter belonged to Sir Francis Scawen Blunt, whose family founded the famous Crabbet Stud of Arab horses. The rider's position was intended to give security over fences on the hunting field.

The Crossroads, 1883, by Henri de Toulouse-Lautrec.

In 1753 John Hawkesworth wrote that the 'same degree of pain in both subjects, is in the same degree an evil',[7] and the conspicuous beating of horses on the road led to condemnation from artists and poets, including William Hogarth and George Stubbs. Thomas Gooch published a series of aquatints called 'The Life and Death of a Race-Horse' in 1783 which drew attention to the suffering of racehorses sold into a working life after their usefulness was over. Criticism rose too over fashionable practices such as tail-docking, which not only cut through the tail-bone but denied the horse his natural means of protection from flies, along with cropping the horse's ears and slitting its nostrils. Narratives in verse, pictures and prose that traced the dissolution of a horse from an owner's pride and joy to a corpse on a knacker's wagon were popular, with the most famous being *Black Beauty* by Anna Sewell, published in 1877.

Although seen today as a children's book, primarily because of the narrative style and the neatly happy ending beloved of the Victorian novelist, *Black Beauty* was written with adults in mind. After a very small first print run of 100 copies, it became a runaway success. In Britain it was endorsed by the Society for the Prevention of Cruelty to Animals, which had formed in 1824 and would go on to become today's RSPCA, while in America it

Design for an elegant barouche coach, 'for use in London and Paris', built *c.* 1880 by Williams of Long Acre, Covent Garden, an area of London well-known for carriage building.

A fatal coach accident, Dublin, 1861.

was distributed to horse-cab drivers with the subtitle 'The Uncle Tom's Cabin of the Horse'. While the first-person narrative by Black Beauty himself highlights the dreadful suffering and overwork of the Victorian city horse, it also shows the hard life of the working family too as, in much of the story, Beauty's owners or handlers suffer along with him.

While Anna Sewell touched the hearts of her readers and raised public consciousness, she came quite late to the debate on suffering. Richard Martin, Member of Parliament for Galway in Ireland, had introduced the first successful legislation to protect animals in 1822. The poet Thomas Hood calls him 'Thou Wilberforce of hacks!', the saviour 'Of whites as well as blacks' along with 'Piebald and dapple-grey' and 'Chestnut and bay'.[8] Likening the treatment of animals to the treatment of slaves illustrates a significant shift in perception. Martin's legislation addressed the needs of farm animals as well as the working horse, starting a slow turn around in the way human responsibility for animals was viewed legally.

Caring for the
working horse at
the Bide-a-Wee
Home for animals,
New York, c. 1906.

But the demands upon the working horse were changing
with the gradual increase of mechanized transport and
machinery, and there was a glimpse of the future in 1830 when
a race took place just outside Baltimore between two railroad
cars. One was pulled by a steam engine named Tom Thumb
and the other by a horse. The engine's builder, Peter Cooper,
aimed to prove to the Baltimore and Ohio Railroad Company
that their plans to use horses to pull cars were outdated. The
horse-drawn car was left behind at once as Tom Thumb took
the lead. But, like the tortoise and the hare, when a belt on
Tom Thumb broke, the horse kept going steadily and won the
race. Yet a point had been made which would increase in

Pontypridd, South Wales, during the miners' strike of 1910–11.

power and make the horse redundant in many working areas over the following century.

It would be a long time before the horse was entirely released from its labours, however, and, even as steam-engines were replacing horses on the railroads of America, pit-ponies were replacing women and children in British coal-mines. The 1842 Mines Act prohibited women and boys under the age of ten from working underground and pit-ponies, already with a long history of mine-work, took over.

There was tension between animal campaigners, who claimed pit-ponies were routinely treated with harsh cruelty, and miners themselves, who blamed the greedy mine-owners for hardships suffered by ponies and workers alike. Many pit-ponies were held in great affection, partly because, like the miners, they worked in hard and dangerous conditions, often being kept underground throughout their working lives. There were claims of cruelty inflicted by the boys, who being over ten

Bob, still working in a Welsh mine in 1997, aged 23. Suffering from lung damage, he was sold to a scrap dealer, but later rescued by the Pit Pony Sanctuary at Pontypridd.

still worked in the mines, but also stories of dedication and selfless care inspired by the individual character of the ponies. The solid companionship of a pit-pony that had accepted its strange working life with all the adaptability of its kind could comfort fears of deep darkness and danger not easily revealed in front of fellow workers. By 1914 there were 70,000 ponies working in British mines, with Welsh Cobs, Dales and Shetland Ponies being among the popular breeds, owing to their small size and great relative strength. Although the numbers dropped over time, when British coal mines were nationalized in 1947 the National Coal Board found itself owning over 20,000 ponies. It would be the late 1990s before the last pit-pony retired in Britain and the popular breeds were exported widely to work in mines, particularly in America.

However, the areas wherein work and pleasure overlap had established different roles for the horse that developed as its working life changed. While entertainments of many types involving horses have been found throughout the centuries, the circus as understood today developed from the late eighteenth century. Horse-trainer Philip Astley was a former Sergeant Major in the cavalry during the Seven Years War and delighted in giving displays of trick-riding near his more formal riding school in Westminster. He soon realized that performing offered him a better future than training and by 1770 he had a full circus established, using the now familiar round ring, which enabled riders to balance through centrifugal force. As well as horses, there were all the popular performers of today, acrobats, jugglers and eventually clowns, co-opted in from centuries of street and court performances.

In 1782 Astley opened the Amphitheatre Anglois in Paris and started a rivalry with Charles Hughes, who had once been a member of his company. Together with Charles Dibdin, who became famed for a quarrelsome temper, Hughes set up the impressively titled Royal Circus and Equestrian Philharmonic Academy in opposition to Astley in London. A former star rider with Hughes and Dibdin, John Bill Rickets, started a fashion for small, travelling circuses which crisscrossed America in brightly coloured wagons throughout the nineteenth century. The more elaborate three-ringed circus, where acts could run simultaneously, was also developed in America by W. C. Coup, while in Europe, innovative tents and seating arrangements meant that the standards of comfort and performance were as high as in a permanent theatre home. By the end of the nineteenth century the 'big-top' had been born and P. T. Barnum, from small beginnings in 1835, opened 'The Greatest Show on Earth'.

Astley was one of many performers who turned to the stage, theatre or circus as a means of continuing a life on horseback that had begun very differently. 'Buffalo Bill' Cody brought together an impressive number of these riders in his huge *Wild West* shows. Born in 1846, William F. Cody had been a cattle herder, fur trapper and gold miner before he went to work for

SERGEANT "PIGGA-NINNI" DISCHARGING HIS COMPANY AFTER DRILL

UNTYING THE KNOT

PASS IT ON

"FLORA" THE TUB ROLLER.

"NETTLE"

TURNING ROUND

"CUPID" THE GUN FIRER.

"BRAVO" &c.

"BRAVO" AND "BONITO"

"RUB WHITE"

C. CORBOULD

"KITTY"

UNDER TRAINING

the Pony Express and scout for the army after the American Civil War. He was still only 26 when he began his show-business career and within ten years the Buffalo Bill's Wild West Show was born. The show aimed to educate as well as entertain and was innovative in the way it presented the Native Americans as an honourable people who had been driven from their lands, rather than perpetuating the popular image of the blood-thirsty enemy. The cast included sharp-shooter Annie Oakley, cowhands recruited from working ranches, and Chief Sitting Bull and his riders of the Hunkpapa Sioux. Sitting Bull would re-enact his own victory at the Little Bighorn as part of the show and amuse himself by insulting his white audiences so that translators had quickly to adapt what he had said into something acceptable. Sitting Bull was eventually murdered in 1890 on the Sioux reservation at Standing Rock and, during the

Sketches by Alfred C. Corbould from a show at the Royal Aquarium, Westminster, in *The Illustrated London News*, 1883.

Poster for *Buffalo Bill's Wild West Show*: 'never twice alike and full of daredevil fun', c. 1910.

Members of Buffalo Bill's show change their horses for a Manchester tram in 1887.

chaos surrounding the attack upon him, the horse given to him by Buffalo Bill went into its performance routine, rearing up on its hind legs to strike high in the air with its front hooves.

Buffalo Bill's Wild West Show famously performed for Queen Victoria in London in 1887, a 200-strong travelling show, which included 97 Oglala Sioux, 180 horses and 18 buffalo. The show was so popular in Manchester that it remained resident for five months and street names such as Dakota Avenue and Buffalo Court remain today to commemorate the visit. By 1891 *Buffalo Bill's Wild West and Congress of Rough Riders of the World* included genuine Cossacks from Russia, Mexican vaqueros and Arab horsemen, and the show's publicity stressed that it differed from other circuses because the cast were not performers but genuine horsemen who used their skills everyday in their native countries.

Feats of equestrianism formed the core of the circus from its beginnings, including bareback riding, acrobatics on horseback and trick-riding. Liberty horses that performed without restraint around the low wall of the circus ring were a firm favourite with audiences. The delight in the horse running free, while nevertheless under the guidance and control of the human, is one of the strongest and most contradictory aspects of the human love of horses. A favourite show on the borderline between theatre and circus was an adaptation of a poem by Lord Byron called 'Mazeppa or the Wild Horse of Tartary'. In this drama, a Polish nobleman, Mazeppa, falls in love with Olinska who, in the best tradition of the melodrama, is to marry an evil count. Mazeppa bests the count in a duel, but instead of letting them both leave unharmed, the count sets his men on Mazeppa. They strip him close to naked and strap him to an untamed horse, which, by a stroke of luck, careers back to Mazeppa's home. Mazeppa was a role famously played by actresses, including Adah Isaacs Menken, who appeared scandalously in the 1860s wearing only a flesh coloured body-stocking to represent the nobleman's disarray. The highly trained 'untamed' circus horse would actually carry Mazeppa rather slowly while the scenery of mountains and rivers moved behind to indicate breakneck speed.

Mazeppa gets carried away. A print by Currier & Ives, 1846.

This fascination with the wild horse, which is really a very tame horse, echoes the movements of the stallion carefully modified for the riding house, the very real dangers of the rodeo and perhaps even the race-course. In the controlled risk of encountering the horse, both the rider and the spectator gain something very specific from participating. Humans love to see the horse leap and run, but also to know it can be controlled. The vicarious danger of watching someone else risk their life on the back of a half-tamed creature offers a satisfying spectacle for the audience, while enabling this second-hand experience of speed and danger validates the risks of the rider. When those risks which had once been essential to survival, as in the case of Buffalo Bill's cast members, became neatly encompassed within the circus ring, the psychological impact on the rider must have been complex. Validation of risk can hardly have reconciled the Hunkpapa Sioux to the radical changes that turned them and their horses into performers. But to the audiences who thronged to see them, only progress could have created a safe environment in which to experience the dangers of the wild frontier as spectacle rather than threat. The decline of the working horse and the rise in industrialization do not simply run parallel to one another or relate in easily definable ways to the thoughts of the changing world about horses. However, the growth of horse-focused entertainments, alongside the changes that would make obsolete the very skills they celebrated, seems unlikely to have been coincidental.

The social contexts that surround humans paradoxically comfort and imprison. By contact with an animal too large to be treated as a pet or to live within the human home, some small contact with wild nature, whether or not acknowledged or even perceived, is maintained. Abstract concepts of nature had inspired the followers of the Romantic movement from

the late eighteenth century onwards. In William Words-
worth's poems 'The Moonlight Horse' and 'The Idiot Boy', a
horse becomes a means by which the essential qualities of
nature are perceived, trusted in its familiarity yet poised on
the borders between the seen and unseen worlds. The
American transcendentalists of the nineteenth century sought
to experience the unfettered qualities of nature to uncover
truths about the human self. Henry David Thoreau ques-
tioned whether it was possible for mankind to remain in
touch with nature while enjoying the benefits of civilization
and included a hound, a bay horse and a turtle dove among the
metaphorical losses of his life. Yet the cabbie trying to make a
living on the city streets, concerned only that his half-starved

This white stallion
in a landscape
by James Ward
(1769–1859) has
both the realistic
detail and
romantic, mythic
quality that typifies
many of Ward's
paintings of
horses.

Safe wildness – stabled horse from Daniéle Da Meda's *Gaúchos*.

horse should make it to the next stop, could have as little sympathy with such lofty preoccupations as with those earnest campaigners who demanded he treat his horse with consideration. Parallels between the plights of animals and slaves were drawn in the early nineteenth century while, towards its close, socialism and feminism offered models that aimed to free the working animal too. The conflict of ideas equalled the chaos of the streets but all the while, the growth of industry and mechanization progressed.

Seeing an emaciated horse dying on the grimy thoroughfares of a city became shocking during the industrial age, while holding the head of a mettlesome stallion under water to teach him a lesson had been quite acceptable in the early Renaissance. Each illuminates the human relationship with the horse and, ultimately, the experience of the horse is always less important than the humans' view of themselves. In 1946 Bertrand Russell wrote, 'The most important effect of machine production on the imaginative picture of the world is an immense increase in the sense of human power',[9] while Thoreau had seen 'A stereotyped but unconscious despair . . . concealed even under what are called the games and amusements of mankind.'[10]

As the breadwinner became less likely to have hooves, the performing horse, re-enacting the Wild West or galloping barely tamed but for years of training, contrived a meeting place between the natural world and that controlled by humans. A head-on-a-stick hobby horse toy had amused children for centuries, but the rocking-horse, lifelike to its mane of real hair, took over during the nineteenth century. Perhaps the ultimate irony was the mechanized toy horse, patented in Washington, DC, in 1867, which offered surely the best of all possible options.

7 The Redundant Horse

> . . . the Report spread of a wonderful Yahoo, that could
> speak like a Houyhnhnm, and seemed in his Words and
> Actions to discover some Glimmerings of Reason.
> Jonathan Swift, *Gulliver's Travels*

In 1972 Southern Television aired a documentary called *Who
Needs Horses?* Now lost, this programme was the product of a
time when new technology seemed to offer the answer to all
questions. The horse was, to the programme maker at least, no
more than a leftover from an earlier time. Desmond Morris
says, 'If a dog is man's best friend, the horse has been man's best
slave.'[1] In the modern world, it seemed to some that the slave
had outlived its usefulness and become a nuisance.

By 1972 my parents had stopped wasting money on toiletries
so I could have a horse. In fact, my sister and I had three by
then, two of which have died only recently, well into their thir-
ties. I spent my time cycling back and forth with my friends to
our ponies, saddles on the handlebars of our bikes ruining the
gears and causing more falls than we ever had from riding. The
idea that no one needed horses was nonsense to us. But 1970s
Britain was difficult for horses in many ways. Rented grazing
was hard to find because shod hooves damage land, making
horses unpopular. Horses lived on odd pieces of land, abandoned golf courses, football fields, allotments and plots
earmarked for building. The rush of horse-owners to buy small-
holdings was only just beginning and, during one difficult
winter towards the end of the decade, horses were given away
and even put down because hay was impossible to come by.

Horses had left the working environment and joined the leisure industry, but just where they fitted in was not entirely clear.

But the 1970s were also the years of the first oil crises. Even as the last horses were coming in from the fields to be replaced by tractors, pleas were beginning for the preservation of the working horse 'as a reserve against a possible disastrous decline in energy resources'.[2] Prophetic words indeed. Working horses had been gradually overtaken by motorized vehicles since the turn of the nineteenth to twentieth century. In 1890s America the cultivation of prairie land for agriculture needed teams of up to 42 horses and 6 men to handle enormous combine harvesters. By 1940 20 million horses had been replaced by tractors. Farmers serving in the British army in France in the First World War had returned so enthused for the Percheron horse that the breed was introduced to work the English farmland, but almost at once the use of heavy horses began to decline with the increase of mechanization. However, the Second World War brought fuel shortages and a directive from the War Agricultural Committee

A United Dairies horse enjoys a nosebag between stops, 1938.

MOLASSINE MEAL MAKES EVEN AN OLD 'CROCK' GO!

CABBY:— "YOU BLOOMIN'
TAXICABS AINT IN IT WITH MY
OLD MARE, WHEN SHE'S 'AD A FEED OF MOLASSINE MEAL!"

The right fuel helps this 'old crock' keep up with the modern competition in 1909.

in 1943 dictated that no tractor should be used for a load a horse could pull. Farmers in areas such as the fenlands of Lincolnshire seemed justified in their reluctance to part with their horses but the future had shown its colours.

Farming had remained, and remains, unmechanized in many parts of the world where the costs of changing to the new technology were prohibitive or horses were too rooted in local

Changing modes of travel in Tokyo, by Yoshitora Utagawa *fl.* 1850–70.

This tearful horse is being carried in a car to the Museum of Natural History by other animals on cars and bicycles. Magazine illustration by J. S. Pughe, 1899.

culture to be easily left behind, but the pressures of economic growth meant that any who could manage the outlay were able to meet the new demands of international agribusiness, a term first used in the 1950s. As food production became a vast global industry, working with horses was simply no longer practical. The development of technology also meant that farmers in difficult terrain, such as the dryland areas of South Australia, could make a viable living for the first time. Machines could accomplish what horses had never been able to, and they could be left unattended and refuelled when convenient. But they didn't offer companionship, any sense of joint effort or the same sense of skilled craft and many farmers who had experienced both hankered for the horse, even as others celebrated the tractor.

However, while the working horse appeared to have been made redundant, a completely new 'role' emerged with the rise of film and television. Between the wars, the Saturday afternoon Western film at the movie theatre or the local 'flea-pit' was an important feature of the week for children on both sides of

Combine harvester and thresher pulled by a large team of horses in Oregon, c. 1903.

the Atlantic. The cultural icon provided by the Western film genre crossed nations to become known worldwide. While the association of white with good and black with evil has a strong tradition, for sheer practicality in black and white film, assigning the hero a white horse to match his pure heart and the villain a black to reflect his evil depths made them easier to identify on a monochrome background. As cowboys galloped across the silver screen from the earliest days of the silent movie, adventure and daring came closer than ever before. Genuine danger was involved for those taking part and a rope called a running-W was used to pull horses off their feet for crashing falls, leading to a long controversy over the use of horses in film. Accusations of mistreatment and concerns over welfare were common, in sharp conflict with the message offered to the innocence of the average viewer, child or adult, that an honest man on a valued horse could always save the day.

Horses began as part of the supporting cast, contextualizing the story and its characters, but this changed as they gathered their own following, taking on some of the attributes seen by wise or reliable horses in myth. Tom Mix in the 1920s was the first of the cowboy-actor stars to give his horse, Tony, billing, followed by Gene Autry, and his horse, Champion, who had been Tony's double before following an independent career. In many instances where films or television series were built around a horse–human partnership, the horse became equally or even more famous than its human companion. Tony starred in two films of his own, *Just Tony* (1922) and *Oh You Tony* (1924) while Champion's career spanned around 20 years, with films, a series of comic books and a television series to his credit. However, heroes never get old, so more than one horse often played a single role which crossed from the screen into real life. There were two Tonys and three Champions, each with several doubles, and Raider, the horse belonging to the Durango Kid, was played by 33 different horses. However, Ken Maynard's horse, Tarzan, was a one-off and often received better reviews than his human co-stars. He danced, leapt off cliffs and saved

Trigger, the ultimate four-legged friend, with Roy Rogers.

his rider from imminent danger regularly between 1923 and his death in 1940, struggling a little in his last years to keep up with his own athletic standard as he and Maynard got older.

The palomino Trigger was one of the most famous film horses, appearing with Roy Rogers, the 'Singing Cowboy', whose career was built around his relationship with his 'four-legged friend'. Trigger showed all the characteristics of the ideal human companion combined with the undemanding acceptance and physical potential of an animal. Being 'honest and faithful right up the end' meant that he undoubtedly never let Rogers down. He also played tricks, covered his rider with a blanket, danced, counted and excelled in chase scenes, speed and surefootedness being two of his natural talents. Like Trigger, Champion, the 'Wonder Horse of the World', as star of his own 26-episode series in 1955–6, behaved like a human throughout, fetching help, fighting crime and standing *in loco parentis* for the young hero, Ricky North. But he also ran wild, his mane and tail blowing in the prairie winds, even as he let Ricky ride him, retaining his horse spirit while assimilating useful human attributes.

However, the most psychologically conflicted horse was Mr Ed, who also starred in his own television series from 1961 to 1966. Mr Ed, the talking horse, wanted to do everything a human could do. He played chess, joined the Peace Corps, needed glasses, used the telephone and aspired to being an artist. Despite his human friend Wilbur's attempts at persuading him to just be a horse, the series ended with Mr Ed planning to train as a doctor. Underneath the comedy, the horse that wanted to be human and the human that wanted him to remain a horse offered a complex commentary on the long equine–human relationship. While a horse that could provide conversation, go for help and run errands would seem to offer everything a human could want, unless it was also content to return to the field and

Mr Ed with Wilbur and his very patient wife Carol, played by Alan Young and Connie Hines.

make no additional demands in return, then we might not be so pleased with it. Champion always went back to the wild once the crisis was over but Mr Ed was so deeply assimilated that his garage-stable was simply no longer enough for him.

With changing times, a role for adventurous girls arose with *Ferien in Lippizza*, a German series from 1965, which was released with an English language soundtrack in 1968 as *The White Horses*. The relationship between Julia, a fifteen-year-old, and Boris, a Lipizzaner horse, forms the background for the series of twelve one-hour stories which was repeated during the 1970s. The huge success of this series in Britain embedded it into the cultural identity of a generation. The popular theme tune, 'On White Horses', sung by Jackie Lee, has since been covered by several other singers and in 2003 was voted the greatest theme tune in television history. Its most recent revival was in a television advert in 2006 to sell children's shoes, featuring a little girl playing on her imaginary horse. While the little girl or any child watching was too young by at least 30 years to remember the song, for the parents or grandparents buying her shoes, it melted away those years very effectively.

In the 1970s more gritty themes, such as those of the *Follyfoot* series, based on books by Monica Dickens, aligned keeping horses with teenage independence and the problems of adolescence. Another famous theme tune, 'The Lightning Tree', and the setting of a rescue home for horses underpinned the themes of recovery and renewal. Storylines covered cruelty, appearances and peer-pressure, while the main character, Dora, was a modern female role-model in a duffle-coat and wellies, rejecting her privileged but emotionally distant upbringing. The horses at Follyfoot Farm were rejects and misfits like the characters, showing their true colours in spite of being misunderstood. Although they did not behave in a human way, like earlier horse-stars, they nevertheless offered foils to human experience. The horse in the imagination of the writer, the filmmaker and their audiences retains all the adaptability that has enabled its survival in the real and changing world.

Those with foresight in the changing world saw that there was still an important part for the horse to play in shaping future generations who understood responsibility and had the skills of balance, co-ordination and concentration, as well as a healthy appreciation of outdoor activity. So in 1928 the Pony Club was established in England to encourage children to learn to ride. It grew to become the largest association of riders in the world, covering around 20 countries with over 110,000 members. Riding clubs and societies for every equestrian activity imaginable exist today and sports involving horses range from racing through endurance riding to carriage driving, gymkhana and eventing. The list is endless and growing, with most sports being

international. With changes in the nature of human work meaning more free time and disposable income, riding as purely a leisure activity is today accessible to a wider group of people than ever before and, in 2004, over a million horses were estimated to be in private rather than commercial ownership in Britain alone.

Equestrian sport has a high mobility so the more formal English-style riding is popular in America while the relaxed approached of Western riding, with a longer leg position and a saddle familiar to non-riders from western films, has a huge following in Germany and a growing profile in France and the UK. The French sport of Le Trec, combining skills of orienteering, trail riding and basic dressage, has left its native country too, to develop internationally.

The chance to enjoy horses is available at the most simple or complicated level on a horse that costs a few hundred pounds to one that costs hundreds of thousands of pounds. Riding has also become a huge spectator sport both live and on television, with equestrian competitions featuring in programmes on sport and having dedicated channels and commentators. Today, we can enjoy the skills and beauty of the horse without ever leaving our seat in front of the television, though whether that is a positive development is debateable.

In fine art the horse has long provided a means of parading wealth and status. Statues and paintings of kings and other notable figures on horses have kept their power in the public eye long beyond the extent of their lives. From the larger-than-life horse that carries Marcus Aurelius through the vigorous muscular horses of Peter Paul Rubens to the elegant Thoroughbreds of George Stubbs, horses have revealed their riders and owners more than they have ever been revealed themselves. With the changing perceptions of the twentieth century, the German artist Franz Marc, who was killed at Verdun in 1916, wanted to

The richly coloured horses of the German Expressionist painter Franz Marc (1880–1916).

179

understand the perspective of the horse, seeing loss and poverty in conventions that prioritized the human view of the world. Horses feature prominently in his work, with *The Large Blue Horses, The Small Blue Horses, The Red Horses* (1911), *The Small Yellow Horses* (1912), all combining primary colours with rounded shapes and an essential quality that captures the nature of the horse. The Italian sculptor Marino Marini drew on the past traditions of Etruscan art in his contemporary imagery of horse and rider while, in 1969, Jannis Kounellis displayed twelve live horses in a vast underground garage in Rome arranged by the Galleria L'Attico in a comment on the manipulative nature of traditional imagery, in its aim of interpreting and presenting the subject in ways influenced by social perspectives.

The distance of the horse from the post-industrial world has made it no less intriguing to the artist and the rise of popular

Artist's impression of Mark Wallinger's proposal for a hill-top landmark near Ebbsfleet in Kent.

culture has meant that even those without the slightest interest in horses remain surrounded by them. Associations with freedom and safe wildness touch humans at a deep level so that any advert, film or story including a horse is destined for success. Black horses are popular in advertising, no longer associated with evil but instead, for example in the UK Lloyds Bank adverts and the rearing black horse of Ferrari, defining sleekness, efficiency and power from financial to mechanical. Cute hairy ponies, typified in the cartoons of Norman Thelwell, offer links to idylls of childhood, while the success of the American trainer and equine behaviorist Monty Roberts and his join-up methods on prime time television, rather than just among horse aficionados, suggests that our desire to retain our connection with the horse survives.

Horses remain central in film and television and, following on from post-war attempts to re-establish a cosy normal and the gutsy themes of the 1970s, in the 1980s leaving hoofprints in space was the next logical step. The rather angry gun-toting 'Equestroid' Thirty/Thirty was an essential companion to Marshall Bravestarr on the futuristic planet, New Texas, in a 1986 animated space Western for children while, with that final frontier to cross, Jean-Luc Picard revealed that he always carried a saddle aboard the *Enterprise*, ready for an inter-galactic horse to present itself. My Little Pony, a popular toy since 1981, continues to develop its range into computer games and an online magazine, taking horses into a rather girly cyberspace. Like Champion the Wonder Horse and Trigger, My Little Pony and Thirty/Thirty offer the idea of horses but give them humanizing features. While for My Little Pony, those features are flowers, hair-ribbons and curly eyelashes, Thirty/Thirty can talk, shoot and change from quadruped to humanoid form. Horses remain a touchstone even in the age of space exploration and web presence, but identification with humans cancels out the physicality of a large animal with a distinctive smell that sheds its coat, sweats copiously as part of its body's cooling system and produces prodigious amounts of dung. Conversely, when Jean-Luc Picard and James T. Kirk finally meet in *Star Trek: Generations* (1994), it is on horseback that the two captains have the discussion that leads to Kirk making the ultimate sacrifice. This is not his life, which he has offered on many occasions previously, but his dream of peaceful retirement on a horse-ranch. Getting back to his earthy roots becomes the ultimate longing even for the space explorer.

It seems to be an expectation among filmmakers that their audiences will not know one horse from another, so it is not unusual for a horse's tack, colour or even breed to change between shots. Horses also tend to be very talkative in film, indulging in far more squealing, neighing and whinnying than they ever do in real life. Interestingly, when horses are depicted in films about war, during which it is widely recorded that they called out in fear or pain, they tend to be silent. But on screen, as so often in art, it is rarely required that horses reveal the nature of horses. Their part is to reveal the nature of the hero, which may explain why the general equine tendency towards flatulence never features in film. This aside, the sense that horses and humans should connect in physical, spiritual and instinctive ways if we are to be whole arises repeatedly. It may be that those toys and films which play down the physicality of the horse reveal the most about our human desire for contact in a form we can understand but have become distanced from.

The presence of the horse in film, television and advertising draws upon its traditional and ongoing ceremonial role. Horses have been used to glorify the monarch or the elite throughout history, which at its simplest level derives from the elevation of the rider over the pedestrian. The horse, expensive to purchase and keep, is a natural stage for the display of skill, ability and wealth. It is still common for riding to be perceived as an elitist activity, although the modern owner comes from many backgrounds. However, horses raise the ceremonial level of royal weddings, military displays and traditional events around the world. The link between the British monarchy and the horse has a long history. Queen Elizabeth II's love of horses is well known, while Prince Philip and Princess Anne and her daughter, Zara Phillips, have kept the British royal equestrian competition profile high since the 1960s. In endurance riding, members of the royal families of Bahrain and Malaysia are among those who distinguish themselves world-wide and act as hosts for key international events. By a strange irony, the means by which elitism was once established, the ability to own and ride fine horses, becomes the humanizing vehicle by which the monarch joins the people.

The traditional high profile alongside its new role in film and the media makes the horse a well-loved and popular animal. It can also make it the deliberate target of violence and the centre of controversy. A series of attacks on horses took place, primarily in the south of England, between 1991 and 1993, many appearing sexual in nature and resulting in the deaths of an unborn foal and an adult horse. While the distress of the owners of the horses involved and the fears of the wider horse-owning community were understandable, the outrage among the media, police force and general public suggested that a sense of wider responsibility had been affronted. Yet horses fill animal welfare centres, horse fairs are widely accepted as places where horses are treated roughly and a great many equestrian sports have aspects that cause controversy. In Australia and America the future of feral horses is at the heart of fierce debate. Horses are used in labora-

Before and after photos of Bahir, a horse rescued and nursed back to health by the International League for the Protection of Horses, which has been active since 1928.

An ILPH farrier working on an international welfare project.

tory experiments and the drugs industry and suffer from live export for the meat-market. The visibility of all these issues is raised with the accessibility of material via the Internet, where, inevitably, pornography involving horses is not unknown.

The tension between the horse as victim and property underlies most of these areas. No matter how much we value the idea of the horse as wild and free, today that is simply an idea. The horse is a domesticated animal, for which we have made ourselves responsible in every way. Our relationship with it is enormously complicated by the anthropocentric tradition that flourishes even as many other grand narratives of the past are rejected. Gifford Pinchot, an early pioneer of conservation, said 'Conservation means the wise use of the earth and its resources for the lasting good of man.'[3] This example of doing the right thing for the wrong reason hinges on a view of human superiority that has shaped the world.

It is not, however, the only view available or even as widely accepted as it may seem. The idea that animals are 'other-than-human' rather than 'non-human' has a long history among many indigenous peoples. Within animistic traditions everything has a soul and equal value in the universe. Western tradition, however, was shaped by the biblical directive that gives humans, more specifically men, dominion over the earth. A correct interpretation of that responsibility has been argued back and forth endlessly and the debate is ongoing. However, in the seventeenth century the philosopher René Descartes argued that animals were governed by instinct, not rational thought, which made them little more than machines. Only rational, therefore human, creatures could have an immortal soul, which meant that humans were 'absolved from any suspicion of crime whenever they kill or eat animals'.[4] While this view was only part of a long debate over the nature and purpose of animals and has never gone unchallenged, it has been nonetheless pervasive. Much of the mistreatment of animals stems from Cartesian logic, which still lurks behind the argument that animal pain is less than human pain, despite any modern spin put upon it. If animals are essentially less than humans, a great burden is raised from the human conscience. But the human urge to colonize extends into the desire to shape, enclose and train so even the concern for suffering can be dominated by a sense that personal or national property has been encroached upon. As the ethics of this human-centred approach come more under scrutiny than ever before, our relationship with the horse, being so central, remains at the heart of one of the most immediate challenges to our cultural traditions.

A ploughing competition in North Wales, 2006.

This gentle Ardennais horse makes a steady logging companion for work with minimum environmental impact.

While the treatment of horses reached a crisis point when they were dying on the streets of London in the late 1800s, their almost complete disappearance from the working role 50 years later was an equal imbalance. There are those who believe it is unethical to ride a horse at all, let alone put it to work, but returning it to the wild is hardly an option and a fat bored horse standing in a field is unlikely to be more content than one that has some occupation. Today societies and individuals who have worked to ensure the survival of working breeds that had become endangered are seeing the fruits of their effort. While horses cannot compete with heavy mechanization, they can work in more subtle ways that may be complementary. Eco-friendly logging and harvesting is possible with horses, which can work between the trees and do less damage to the forest floor. They also avoid the inherent danger in the vast potential of sophisticated machinery for causing rapid irrevocable damage. The reintroduction of the Przewalski's Horse into its natural home was only the beginning of a consideration of the value of the grazing horse in land management. Since the 1990s Przewalski's Horses have been introduced into Britain, Germany, Austria, Hungary, France, Ukraine, China, Mongolia, Uzbekistan and Kazakhstan, in reserves or semi-reserves, with varying levels of monitoring and support. Domesticated breeds play a similar part, for example in Wales, where, on the small island of Anglesey, Shetland ponies graze the heathland and on the slopes of Cader Idris, Welsh Mountain ponies enjoy willow scrub. In both instances the ponies have been deliberately introduced into ecology programmes as their natural grazing habits control the vigorous plants that can overwhelm rare species such as bog asphodel and marsh orchid.

Horses can also take people into remote places while leaving little evidence of their journey. Environmental tourism projects open up a world on horseback and today it is possible to ride landscapes as varied as the steppes of Mongolia, the bushveld of Mozambique and the volcanic drama of Iceland on trail rides that are specifically designed to cause as little environmental impact as possible. Simply being with horses is widely acknowledged as a valuable experience and arising from this are projects across a wide range of countries which involve horses in therapeutic work with children and adults who have suffered trauma or illness. The horse has a

Eco-friendly Przewalski's Horses in Clocaenog Forest, North Wales, helping to clear the site of an ancient settlement.

Trekking in Nepal with ponies.

role in the newly thoughtful world which is far more suited to its own nature than riding it into war could ever be.

Over the centuries, witnesses to superb riding and those who set out to train others constantly use the image of the one-ness of mind and body between rider and horse to express an ideal. These Centaur longings find expression today in the desire to work from the perspective of the horse and the study of the horse's mind is perhaps a natural result of the modern interest in psychology. The Equine Behaviour Forum at the University of Glasgow is an example of the academic and the interested owner coming together to develop theory through practice. By shared observation and experience new approaches to the horse are developing all the time, with a focus on the importance of the herd environment to mental well-being and a developing interest in the value of keeping horses 'barefoot' where possible. Non-resistance training seeks to approach the horse from a position of understanding its natural psychology

Maintaining an old tradition in Brittany in 2007: a travellers' caravan . . .

and is a vast and growing movement. The concept of the horse whisperer is founded in this method and while it is a wonderfully romantic image to think that we can soothe the troubled horse by whispering to it, the method in practice is not mystical

. . . and their horse.

or even mysterious. Perhaps the strength in the image of the horse whisperer is that it is quiet and thoughtful, running counter to the idea that a fierce bit and a bigger stick can sort out a horse perceived as just being difficult. But talking horses and humans who understand them appear in myth and fiction so frequently that this longing to communicate seems to be of exceptional importance. When a 'one-sided or hierarchical dynamic gives way to reciprocity between ourselves and other species',[5] new possibilities can open up for both parties.

When we look at the horse, what we see depends entirely on our own perceptions of the world, which are in themselves always changing. Ultimately the horse reflects those perceptions, both in the way we treat it and the way it responds. The mettlesome creatures ready to curvet out of the picture frame in

JR and Wayne Delbeke tackle the infamous Cougar Rock on the Tevis Cup Endurance Ride. Note how little the horse has on his head and the shared focus – a great example of partnership between horse and rider.

In 1560, Sir Thomas Blundeville would have said the horse in this image sequence is rolling because he was born under the sign of Leo. Today we would say it is to keep his coat healthy.

This Appaloosa is the loudly spotted young horse in the family pony montage on page 11, showing how the distinctive coat patterns of the breed may change over the years.

portraits of the nobility dating from the Renaissance reflect an understanding of the self as something dangerously wild, to be kept in check, bridled, managed and harnessed: all common terms derived from handling horses. But they also illustrate the way in which a horse responds to that approach. It is not uncommon to hear that a rider has an 'electric seat', which means that any horse they mount becomes agitated and unsettled. There is something in the involuntary physical and mental approach of that rider that transmits to the acute sensitivities of the horse at once. Any competent rider knows that having the presence of mind, and often the courage, to look in the opposite direction to something that has alarmed the horse they are riding offers the best chance that their mount will decide it is not so very frightening after all. A quietly unconcerned rider can do more to calm a restive horse than any amount of training.

The horses in several famous portraits of Charles I have received comment for being remarkably self-contained. They

Quiet rider, quiet horse: the calm authority of a monarch mirrored in his dun stallion in *Charles I on Horseback* by Anthony van Dyck, c. 1637–8.

appear 'not exactly like a horse' and might indeed be 'admitted into a drawing-room without offence',[6] because they mirror the monarch's calm and meditative demeanour, a fact often seen by art critics as an entirely metaphorical feature of the work. The metaphorical undercurrents undoubtedly influence our reading of the paintings, but their accuracy as an observation of the horse's nature is underestimated. They are, in fact, exactly like a horse and thus reveal more about Charles I than any metaphor ever could. In film, horses are trained to perform in ways that reflect the nature of the hero, but their natural response to the actor playing the role might be quite different, which explains why those horse–rider combinations that have become most famous are based on real-life partnerships. What we see is very much what we get and this remains as true today as it ever did. The way in which we gain a reflection of ourselves through the horse that pleases us reveals us at our most psychologically naked.

The horse is written so intrinsically into human history that attempting to identify key aspects of its role becomes an exercise in leaving out rather than including. Without the horse, humans would be different and for every example I have chosen to illustrate that fact, there are countless others. There are myths, breeds, cultures, wars, roles both ancient and modern enough for a great many books. But no matter how much evidence there is to consider, how much material to be sifted, the outcome remains the same: at a level whose nature we can debate endlessly, the horse matters to us.

While we do that debating, the horse just 'scratches its innocent behind on a tree',[7] and this is perhaps its ultimate appeal. Through all the exploration, adventure and conflict of human life, the horse keeps on being a horse, adapting admirably, teaching us to be good leaders, warning us not to take too much for granted. We can romanticize it very easily for its association with wildness and freedom or take out our frustrations upon it because that appeals to some dark side of ourselves that we are unlikely to acknowledge. But it seems that in one way or another we are unwilling, or even unable, to do without it and when we approach it in that knowledge, we begin to fully realize the potential of our long shared history.

Timeline of the Horse

20–1 MILLION BC	C. 4,000–3,000 BC	C. 350 BC	C. 100 BC
Equus caballus evolves from the long slow processes of natural selection that shape the many branches of the equids	Horses are hunted and kept as domestic stock in southern Ukraine, north Kazakhstan and western Siberia	The Greek cavalry commander Xenophon writes a treatise on training a horse	The White Horse of Uffington is carved into the chalk hills of the Berkshire Downs

1714	1769–70	1806	1821
Flying Childers is foaled; he will go down in racing history as the fastest horse ever born	Eclipse, great-great-grandson of the Darley Arabian, wins 18 races and is never beaten on the track	The quality of the Appaloosa horses bred by the Nez-Perce is noticed by Lewis and Clarke	The term 'English Thoroughbred' is first used

1914–18	1928	1943
The British Army loses 500,000 horses in World War 1	The Pony Club is established in England	In Britain horses replace tractors during the Second World War to save fuel

c. AD 500–900	1448	1532	1541
Celtic settlements in north Britain breed horses to type for riding, draught and agriculture	At the Battle of Caravaggio 10,000 horses are killed	Federigo Grisone opens his riding school in Naples	Mounted Spanish conquistadors reach the plains of present-day Kansas

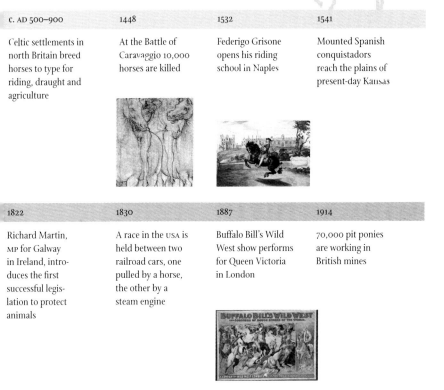

1822	1830	1887	1914
Richard Martin, MP for Galway in Ireland, introduces the first successful legislation to protect animals	A race in the USA is held between two railroad cars, one pulled by a horse, the other by a steam engine	Buffalo Bill's Wild West show performs for Queen Victoria in London	70,000 pit ponies are working in British mines

1961–6		1998	2004
Mr Ed, the TV talking horse, ponders the best way to become more like a human		The number of Przewalski's Horses reintroduced to their native home reaches 60	Over a million horses are estimated to be in private rather than commercial ownership in Britain alone

References

INTRODUCTION

1 William Cavendish, *A New Method, and Extraordinary Invention, to Dress Horses* (London, 1667), sig. (c)2.
2 William Holt, *Ride a White Horse: A 9000 Mile Equestrian Odyssey* (London, 2001), pp. 7–10.
3 Jeremy James, *Vagabond* (London, 1991), pp. 145, 182.
4 Lucy Rees, *The Horse's Mind* (London, 1991), pp. 198–9.

1 EOHIPPUS TO EQUUS

1 Gordon Rattray Taylor, *The Great Evolution Mystery* (London, 1983), p. 156.
2 David Rains Wallace, *Beasts of Eden* (New Haven, CT, 2005), p. 50.
3 Quoted in Hope Ryden, *America's Last Wild Horses* (Guildford, CT, 2005), p. 21.
4 Rudyard Kipling, *The Just So Stories* (London, 1989), p. 80.

2 PEGASUS, EPONA AND DEMETER'S FOALS

1 Howard Schwartz, *Miriam's Tambourine: Jewish Folktales from Around the World* (Oxford, 1988), p. 144.
2 Dane and Mary Roberts, *The Navajo Indians* (Boston, 1930), p. 1.
3 J.R.R. Tolkien, *The Two Towers* (London, 1969), p. 547.
4 Quoted in Katherine Briggs, *A Dictionary of Fairies* (Harmondsworth, 1979), p. 148.

5 M. Oldfield Howey, *The Horse in Magic and Myth* (New York, 2002), p. 213.
6 Christopher Marlowe, *Dr Faustus*, ed. John Butcher (London, 2001), Act 4 sc. v, lines 11–13.
7 Thomas Blundeville, *A Newe Booke containing the Arte of Ryding and breaking greate Horses* (London, 1560), p. A.iiiv.
8 William Cavendish, *A General System of Horsemanship: Volume 1* (London, 1743), p. 20.
9 Schwartz, *Miriam's Tambourine*, p. 367.

3 THE MAN-MADE HORSE

1 Elwyn Hartley Edwards, *The Encyclopaedia of the Horse* (London, 1994), p. 169.
2 Hossein Amirsadeghi and Peter Upton, *The Arabian Horse: History, Mystery and Magic* (London, 1998), p. 7.
3 Thomas Blundeville, A *Newe Booke containing the Arte of Ryding and breaking greate Horses* (London, 1560), p. 6.
4 William Cavendish, *A New Method, and Extraordinary Invention, to Dress Horses* (London, 1667), pp. 72–3.
5 Charles Darwin, *The Origin of Species* (London: 1872), pp. 22–3, 31.

4 RIDING INTO HISTORY

1 'The Horse in Mongolian Culture', *The American Museum of Natural History Science Bulletin*, Bio Feature : 'The Last Wild Horse,' www.amnh.org/sciencebulletins/[accessed June 2008].
2 Thomas Blundeville, *A Newe Booke containing the Arte of Ryding and breaking greate Horses* (London, 1560), pp. C.i–C ii, Aiiii, Ciiii.
3 Antoine de Pluvinel, *The Maneige Royal,* trans. Hilda Nelson (London, 1989), pp. 24–5.
4 Philip Sidney, *Defence of Poesie, Astrophil and Stella and Other Writings*, ed. Elizabeth Porges Watson (London, 1997), p. 83.
5 Thomas Elyot, *The Boke called the Governour* (London, n.d.), Book XVII.

6 William Cavendish, *A New Method, and Extraordinary Invention, to Dress Horses* (London, 1667), sig. (c).

7 Ibid., p. 4.

8 Quoted in Alvin M. Josephy, Jr, *500 Nations: An Illustrated History of North American Indians* (London and New York, 1995), p. 359.

9 Quoted ibid., p. 362.

10 Quoted in Clyde A. Milner II et al., eds, *The West: An Illustrated History* (New York and Oxford, 1994), pp. 275, 273.

11 Aldo Sessa, *Gauchos* (Cologne, 2001), p. 7.

12 Xenophon, *The Art of Horsemanship*, trans. M. H. Morgan (London, 1993), p. 40.

13 Zane Grey, *Wildfire* (London, n.d.), p. 9.

14 James Fillis, *Breaking and Riding with Military Commentaries*, trans. M. H. Hayes (Kingston, n.d.), p. 24.

15 Melissa Holbrook Pierson, *Dark Horses, and Black Beauties* (New York and London, 2001), pp. 60–61.

5 INTO THE VALLEY OF DEATH

1 George Bernard Shaw, *Arms and the Man* (Edinburgh, 1901), Act 1.

2 Quoted in Ian Fletcher, *Galloping at Everything: The British Cavalry in the Peninsular War and at Waterloo 1808–15* (Staplehurst, 1999), p. 13.

3 Quoted in Ann Hyland, *The Warhorse 1250–1600* (Stroud, 1998), p. 70.

4 Quoted in Greg Hetherington, *Britain and the Great War: A Study in Depth* (London, 1998), p. 46.

5 Quoted in Richard Holmes, *Riding the Retreat, Mons to Marne: 1914 Revisited* (London, 2007), pp. 186–7.

6 Quoted in ibid., pp. 247–8.

7 Quoted in Hetherington, *Britain and the Great War*, p. 46.

8 Holmes, *Riding the Retreat*, pp. 292–3.

9 Dorothy Brooke, letter to the *Morning Post*, 1931: www.thebrooke.org/content.asp?id=15567 [accessed 23 June 2008].

6 FROM BREADWINNER TO PERFORMER

1 Thomas Hardy, *Tess of the D'Urbervilles* (Harmondsworth, 1978), pp. 70–73.
2 Charles Dickens, *The Pickwick Papers* (London, 1986), p. 289.
3 Quoted in Russell Lyon, *The Quest for the Original Horse Whisperers* (Edinburgh, 2003), p. 10.
4 In Tom Regan and Andrew Linzey, eds, *Song of Creation: An Anthology of Poems in Praise of Animals* (Basingstoke, 1988), p. 59.
5 W. J. Gordon, *The Horse World of London* (London, 1893), p. 10.
6 Jacques de Solleysel, *Le Parfait Mareschal or Complete Farrier*, trans. William Hope, and including *The Compleat Horseman: A Compendious Treatise of the Art of Riding, collected from the best Modern Writers on the Subject* (Edinburgh, 1696), p. 92; William Cavendish, *A New Method, and Extraordinary Invention, to Dress Horses* (London, 1667), pp. 132–3.
7 Quoted in Malcolm Warner and Robin Blake, *Stubbs and the Horse* (New Haven and London, 2005), p. 9.
8 In Regan and Linzey, eds, *Song of Creation*, p. 83.
9 Bertrand Russell, *A History of Western Philosophy* (London, 1983), p. 699.
10 Henry David Thoreau, *Walden: or, Life in the Woods* (New York, 2002), p. 8.

7 THE REDUNDANT HORSE

1 Desmond Morris, *Illustrated Horsewatching* (London, 1997), p. 7.
2 George Ewart Evans, *Horse Power and Magic* (London, 1979), p. 3.
3 Clyde A. Milner II and others, eds, *The West: An Illustrated History* (New York and Oxford, 1994), p. 327.
4 René Descartes, *Meditations and Other Metaphysical Writings*, trans. Desmond M. Clarke (London, 2003), p. 175.
5 Adele von Rüst McCormick and Marlena Deborah McCormick, *Horse Sense and the Human Heart* (Florida, 1997), p. 17.

199

6 Quoted in Roy Strong, *Van Dyck: Charles I on Horseback* (New York, 1972), p. 38.

7 From W. H. Auden, 'Musée de Beaux Arts', in *The Oxford Library of English Poetry*, vol. III, ed. John Wain (London, 1988), pp. 401–2.

Bibliography

Alciato, Andrea, *Emblematum liber* (1531), Alciato Project
  <http://mun.ca/alciato/desc.html> [accessed 23 June 2008]
Amirsadeghi, Hossein and Peter Upton, *The Arabian Horse: History,*
  Mystery and Magic (London, 1998)
Aspley, Viola, *Bridleways Through History* (London, 1936)
Berger, John, *About Looking* (New York, 1980)
Bird, Isabella, *A Lady's Life in the Rocky Mountains* (London, 1982)
Blundeville, Thomas, *A Newe Booke containing the Arte of Ryding and*
  breaking greate Horses (London, 1560)
Brown, Dee, *Bury my Heart at Wounded Knee: An Indian History of the*
  American West (London, 1972)
Budiansky, Stephen, *The Nature of Horses: Their Evolution, Intelligence*
  and Behaviour (London, 2004)
Burt, Jonathan, *Animals in Film* (London, 2002)
Cavendish, William, *La Methode Nouvelle et Invention extraordinaire de*
  dresser les Chevaux (Antwerp, 1658)
——, *A New Method, and Extraordinary Invention, to Dress Horses*
  (London, 1667)
——, A *General System of Horsemanship: Volume 1* (London, 1743)
Chasteen, John Charles, *Heroes on Horseback: The Life and Times of the*
  Last Gaucho Caudillos (Albuquerque, NM, 1995)
Clutton-Brock, Juliet, *A Natural History of Domesticated Animals*
  (Cambridge, 1999)
Coolidge, Dane and Mary Roberts, *The Navajo Indians* (Boston, MA,
  1930)

Darwin, Charles, *The Origin of Species* (London, 1872)

——, *The Origin of Species,* ed. J. W. Burrow (London, 1985)

Davies, Sioned and Nerys Ann Jones, eds, *The Horse in Celtic Culture: Medieval Welsh Perspectives* (Cardiff, 1997)

Dent, Anthony, *The Horse Through Fifty Centuries of Civilisation* (London, 1974)

Descartes, René, *Meditations and Other Metaphysical Writings*, trans. Desmond M. Clarke (London, 2003)

Deusen, Kira Van, *Singing Story, Healing Drum: Shamans and Storytellers of Turkic Siberia* (Montreal, 2005)

Edwards, Elwyn Hartley, *The Encyclopaedia of the Horse* (London, 1994)

Edwards, Peter, *The Horse Trade of Tudor and Stuart England* (Cambridge, 2004)

Elyot, Thomas, *The Boke called the Governour* (London, n.d.)

Evans, George Ewart, *Horse Power and Magic* (London, 1979)

Ewers, J. C., *The Horse in Blackfoot Indian Culture* (Washington, 1980)

Fillis, James, *Breaking and Riding with Military Commentaries*, trans. M. H. Hayes (Kingston, n.d.)

Fletcher, Ian, *Galloping at Everything: The British Cavalry in the Peninsular War and at Waterloo 1808–15* (Staplehurst, 1999)

Gordon, W. J., *The Horse World of London* (London, 1893)

Grey, Zane, *Wildfire* (London, n.d.)

Guérinière, François Robichon de la, *School of Horsemanship*, trans. Tracey Boucher (London, 1994)

Hetherton, Greg, *Britain and the Great War: A Study in Depth* (London, 1998)

Holmes, Richard, *Riding the Retreat, Mons to Marne: 1914 Revisited* (London, 2007)

Holt, William, *Ride a White Horse: A 9000 Mile Equestrian Odyssey* (London, 2001)

Howey, M. Oldfield, *The Horse in Magic and Myth* (New York, 2002)

Hyland, Ann, *The Medieval Warhorse from Byzantium to the Crusades* (Stroud, 1994)

——, *The Warhorse, 1250–1600* (Stroud, 1998)

Irwin, Chris, *Horses don't Lie* (London, 2001)

Jackson, Jaime, *Horse Owners Guide to Natural Hoof Care* (Harrison, AR, 2000)

James, Jeremy, *Vagabond* (London, 1991)

——, *The Byerley Turk* (Ludlow, 2005)

Johns, Catherine, *Horses: History, Myth, Art* (London, 2006)

Josephy Jr, Alvin M., *500 Nations: An Illustrated History of North American Indians* (London and New York, 1995)

Kean, Hilda, *Animal Rights: Political and Social Change in Britain since 1800* (London, 1998)

Kingston, W.H.G., *Stories of the Sagacity of Animals* (London, 1879)

Kipling, Rudyard, *The Just So Stories* (London, 1989)

Lawrence, D. H., *Apocalypse* (London, 1931)

Lewis, C. Jack, *White Horse, Black Hat: A Quarter Century on Hollywood's Poverty Row* (Oxford, 2002)

Lewis, Jon E., *The West: The Making of the American West* (Bristol, 1996)

Liedtke, Walter A., *The Royal Horse and Rider: Painting, Sculpture and Horsemanship 1500–1800* (New York, 1989)

Lyon, Russell, *The Quest for the Original Horse Whisperers* (Edinburgh, 2003)

Manning, Aubrey and James Serpell, eds, *Animal & Human Society: Changing Perspectives* (London and New York, 1994)

McCormick, Adele von Rüst and Marlena Deborah, *Horse Sense and the Human Heart* (Deerfield Beach, FL, 1997)

Milner II, Clyde A., Carol A. O'Connor and Martha A. Sandweiss, eds, *The West: an Illustrated History* (New York and Oxford, 1994)

Montaigne, Michel de, *The Complete Essays*, ed. M. A. Screech (London, 1991)

Morris, Desmond, *Illustrated Horsewatching* (London, 1997)

Parelli, Pat, *Natural Horse-Man-Ship* (Colorado, 1993)

Pierson, Melissa Holbrook, *Dark Horses, and Black Beauties* (New York and London, 2001)

Piggot, Stuart, *Wagon, Chariot and Carriage: Symbol and Status in the History of Transport* (London, 1992)

Pluvinel, Antoine de, *The Maneige Royal*, trans. Hilda Nelson

(London, 1989)

Rashid, Mark, *Considering the Horse* (Colorado, 1993)

Rees, Lucy, *The Horse's Mind* (London, 1991)

Regan, Tom and Andrew Linzey, eds, *Song of Creation: An Anthology of Poems in Praise of Animals* (Basingstoke, 1988)

Remington, Frederic, *Frederic Remington's Own West*, ed. Harold McCracken (London, 1960)

Richardson, Bill and Dona, *The Appaloosa* (California, 1969)

Roberts, Monty, *The Man who Listens to Horses* (London, 1997)

Ryden, Hope, *America's Last Wild Horses* (Guildford, CT, 2005)

Sessa, Aldo, *Gauchos* (Cologne, 2001)

Sewell, Anna, *Black Beauty* (London, 1988)

Shaw, George Bernard, *Arms and the Man* (Edinburgh, 1901)

Sidnell, Philip, *Warhorse: Cavalry in Ancient Warfare* (London and New York, 2006)

Solleysel, Jacques de, *Le Parfait Mareschal or Complete Farrier*, trans. William Hope, and including *The Compleat Horseman: A Compendious Treatise of the Art of Riding, collected from the best Modern Writers on the Subject* (Edinburgh, 1696)

Strong, Roy, *Van Dyck: Charles I on Horseback* (New York, 1972)

Schwartz, Howard, *Miriam's Tambourine: Jewish Folktales from Around the World* (Oxford, 1988)

Swift, Jonathan, *Gulliver's Travels*, ed. Albert J. Rivero (New York and London, 2002)

Taylor, Gordon R., *The Great Evolution Mystery* (London, 1983)

Thirsk, Joan, *Horses in Early Modern England: for Service, for Pleasure, for Power* (Reading, 1978)

Thomas, Keith, *Man and the Natural World: Changing Attitudes in England, 1500–1800* (London, 1983)

Thoreau, Henry David, *Walden: or, Life in the Woods* (New York, 2002)

Tolkien, J.R.R. *The Two Towers* (London, 1969)

Toole-Stott, R. S., *Circus and Allied Arts: A World Bibliography 1500–1959* (Derby, 1960)

Unger-Hamilton, Clive, *The Entertainers: A Biographical History of the Stage* (London, 1980)

Walker, Elaine, 'The Duke of Newcastle and the Spanish Horse',
    Andalusian: Dedicated to the Spanish and Portuguese Horse
    (Birmingham, AL, 2003), pp. 21–5
——, 'To Amaze the People with Pleasure and Delight: An Analysis
    of the Horsemanship Manuals of William Cavendish, First Duke
    of Newcastle (1593–1676)', doctoral thesis, University of
    Birmingham, 2005
Wallace, David Rains, *Beasts of Eden* (New Haven, CT, 2005)
Wangford, Hank, *Lost Cowboys* (London, 1995)
Warner, Malcolm and Robin Blake, *Stubbs and the Horse* (New Haven,
    CT, and London, 2005)
White, Raymond E, 'B-Western Horses', *Western Horseman* (CO,
    2001), pp. 54–9
Yates, Roger and Piers Beirne, 'Horse Maiming in the English
    Countryside: Moral Panic, Human Deviance and the Social
    Construction of Victimhood', *Society & Animals Journal of
    Human–Animal Studies*, IX/1 (2001), available at www.psyeta.org/
    sa/sa9.1/yates.shtml (accessed 27 March 2008)
Xenophon, *The Art of Horsemanship*, trans. M. H. Morgan (London,
    1993)

Associations and Websites

THE BROOKE
www.thebrooke.org
A major international welfare organization concerned with all
equines and the people who rely upon them.

ECOLE NATIONALE D'EQUITATION/ LE CADRE NOIR DE SAUMUR
www.cadrenoir.fr
An impressive facility just outside Saumur in the Loire Valley of
France.

EQUINE BEHAVIOUR FORUM
www.gla.ac.uk/external/EBF
An international group open to horse-owners, academics, horse-
industry professionals and anyone with an interest in equine
behaviour.

FOUNDATION FOR THE PRESERVATION AND PROTECTION OF THE
PRZEWALSKI HORSE
www.treemail.nl/takh
Comprehensive information on this last link to the wild past.

H-ANIMAL
www.h-net.org/~animal
An online cross-disciplinary discussion group for scholars
studying animals in human culture.

THE HORSE: A MIRROR OF MAN: PARALLELS IN EARLY HUMAN AND
HORSE MEDICINE
www.nlm.nih.gov/exhibition/horse/gallery.html
National Library of Medicine site, which includes images from
early and rare texts.

INTERNATIONAL LEAGUE FOR THE PROTECTION OF HORSES
www.ilph.org
Offers information on the history of the charity, current campaigns, welfare centres and adoption procedures.

OUVRAGES ANCIEN
www.vet-lyon.fr/22224257/0/fiche___pagelibre/&RH=
1164811905831.
A virtual library of early horsemanship and veterinary texts, usually only found in library special collections and rarely together.

PIT PONY SANCTUARY, PONTYPRIDD, WALES
www.pitponies.co.uk
Registered charity and working smallholding that houses former
pit-ponies. Open for visitors.

PONY CLUB
www.pcuk.org
The largest association of riders in the world. Its informative web-
site offers access to all aspects of its international activities.

SPANISH RIDING SCHOOL OF VIENNA
www.srs.at/index.php?id=353

Acknowledgements

I would like to thank my husband, Bryan Walker, for his encouragement and support, as always, and my son, Christian, for reading several drafts of this book. Thanks also to my friends, Avril de Cambray and Lynda Tudor, for their patience in listening to the history, anecdotes and sometimes distressing facts I've come across in my research. I would also like to thank the large number of people, most of whom I have never met, who have been kind enough to share their photographs, stories and experience with me. In particular, thanks to Mal Hughes of J and M Photography for his technical expertise; Daniéle Da Meda, Ruth Wilbur, the Pit Pony Sanctuary, Magnolia Hotels Ltd, Turfsurfer Ltd, Meyfroidt Photography, Miraval Andalusians, the International League for the Protection of Horses and Llantrisant Folk Club for allowing me to use their photographs; Zoe Skoülding and Trevor Foulds for their time; Iolo Lloyd, Conservation Officer for Clocaenog Forest, for taking me to visit his Przewalski's Horses, Mic Rushden, of Solva Icelandic Horses, and Wayne Delbeke, of Lazy Ed Canabians, who invited me to South Wales and British Columbia to ride their horses. Finally, a mention for all the horses who have shaped my way of thinking over the years, especially JB, Darius, Topaz and the late, great Cuddly Jim, as well as Cadbury and Zebedee, who are not horses, but are vital long-eared members of the herd nevertheless.

Photo Acknowledgements

The author and publishers wish to express their thanks to the below sources of illustrative material and / or permission to reproduce it.

Academy of Sciences, Leningrad: p. 70; Bayerische Staatsgemälde-sammlungen, Munich: p. 178; collection of the author: pp. 17, 31, 67, 75, 78, 98, 101, 114, 132, 134, 137, 140, 144, 153 (top), 154, 155, 160, 171, 187, 189, 191; Cyngor Sir Ceredigion County Council (Ceredigion Archives): p. 138; The British Museum, London (photo The Trustees of the British Museum): p. 113; photo © Dave Collier: p. 48; © Daniéle Da Meda: pp. 19, 111, 166 (all images from her exhibition 'Gauchos'); Foundation for the Preservation and Protection of the Przewalski's Horse: p. 32; photo © Karen Givens / 2008 iStock International Inc.: p. 71; Gulistan Library, Teheran: p. 92; photo Hughes Photography: p. 190; Image*After: pp. 120, 188; photo The International League for the Protection of Horses: p. 184; Islamisches Museum, Berlin: p. 12; Istanbul University Library: p. 60; photo © Randy Harris / Big-StockPhoto: p. 15; J & M Photography: pp. 11, 186; photo William Henry Jackson / Library of Congress, Washington, DC (Prints and Photographs Division, World's Transportation Commission photograph collection): p. 57; Japan Folk Art Museum, Tokyo: p. 30; Leeds City Art Galleries: p. 122; Library of Congress, Washington, DC (Prints and Photographs Division): pp. 13, 80, 104, 108, 116 (British Cartoon Prints Collection), 124, 129, 136 (Cabinet of American Illustration), 141 (Civil War drawing collection), 156, (George Grantham Bain Collection), 157, 161, 163, 170, 172, 173; photo © Llantrisant Folk Club· p. 52; by permis-

Index